ROBERT MANNING STROZIER LIBRARY

TALLAHASSEE, FLORIDA

BRITISH GENERALSHIP
IN THE TWENTIETH CENTURY

Map 3.

BRITISH GENERALSHIP IN THE TWENTIETH CENTURY

by

Major-General E. K. G. Sixsmith

ARMS AND ARMOUR PRESS

Published by
Arms and Armour Press
Lionel Leventhal Limited
677 Finchley Road
Childs Hill
London, N.W.2

© E. K. G. Sixsmith, 1970

All rights reserved
SBN 85368 039 6

To my Regiment
The Cameronians (Scottish Rifles)
1689-1968

Printed by The Central Press (Aberdeen) Ltd.
Belmont Street, Aberdeen

CONTENTS

	page
List of Illustrations	7
List of Maps	8
Introduction	9

CHAPTER 1
  The South African War — 13

CHAPTER 2
  Army Reorganization and the Creation of the Expeditionary Force — 31

CHAPTER 3
  1914: The Encounter Battles — 45

CHAPTER 4
  The Search for a Way Round — 54

CHAPTER 5
  The Western Front: Neuve Chapelle to Passchendaele — 80

CHAPTER 6
  Cambrai — 113

CHAPTER 7
  1918 — 127

CHAPTER 8
  Haig and Hamilton — 147

CHAPTER 9
  Between the Wars — 163

Contents

                                                                                       page

CHAPTER 10
    The Strategy of the Second World War   186

CHAPTER 11
    Tactical Aspects of the War in the Desert   213

CHAPTER 12
    Tactical Aspects of the War in Europe   243

CHAPTER 13
    The War against Japan   265

Epilogue   294
Appendix   297
Index   301

LIST OF ILLUSTRATIONS

Plate		Facing Page
1	H.M. King George V in Flanders before the third Battle of Ypres	96
2	Tank crashing through barbed wire, 1917	97
3	The Hindenburg Line near Arras, 1917	97
4	The Hindenburg Line near Bony, 1918	112
5	Infantry and tanks waiting to attack, 1918	112
6	Canadians training with tanks, 1918	113
7	The Hindenburg Line at Queant, 1918	113
8	Tanks in action on the Western Front, 1918	176
9	Tanks and infantry passing through a captured village near Caumont, 1944	176
10	The Battle of Vimy Ridge, 1917	177
11	British infantry advancing in North Africa, 1942	192
12	Tanks pass through infantry holding positions near Caen, 1944	193

LIST OF MAPS

Map		page
1	North-West Europe	front endpaper
2	The Mediterranean and South-East Europe	back endpaper
3	South Africa	frontispiece
4	The Somme	94
5	Arras and the Artois Battles, 1917	103
6	Third Battle of Ypres, 1917	107
7	Cambrai, 1917	117
8	Gallipoli	150
9	The Far East and the Western Pacific	203
10	The Desert Campaign	221
11	Central Italy	246
12	Malaya	267
13	The Burma Campaign	284

INTRODUCTION

The theme which runs through this book is the apparent conflict between two essential military requirements; on the one hand the will to stand and fight and the habit of unquestioning obedience; on the other the need for initiative, for subtlety and subterfuge, and the necessity to use all possible means of overcoming the will of the enemy at the least cost to oneself. A lack of the resolution inherent in the first was the undoing of Buller in the Boer War, and the double pull is seen in Roberts's conduct of operations. It is, however, in the Great War that the conflict becomes fully apparent. It is wrong to believe that British strategy on the western front was unthinking and blind concentration on the first essential; there were real efforts to resolve the deadlock, but undoubtedly some of their mistakes from 1915 to 1917 were rooted in a failure to reconcile the two conflicting requirements. The cost of the resolution of the problems of the Great War made a deep impression on the minds of those who were to become the commanders in the Second World War; the extent to which they understood the reason and learnt from their mistakes is the subject of the later chapters of the book.

Much of the information about the Great War has been obtained from research into the Maxse Papers (Imperial War Museum) and records in possession of the Royal Artillery Institution. For the later period personal knowledge and experience naturally plays a large part. In particular a knowledge of Field-Marshal Montgomery's approach to military problems comes from my two years at the Staff College, Quetta, during the whole of which time Montgomery was a Chief Instructor. In some ways I should have preferred not to have to write about the Second World War because to do so is inevitably to pass judgment or at least to criticize men under whom I have served and men whom I know to have more experience of command in battle than I have myself. Nevertheless to do so was inevitable if I was to fulfil the theme.

I am indebted to many friends for help in writing the book. Major-General R. C. Money has gone to endless pains to assist me in the chapters on the Great War; Lieutenant-General Sir John Evetts has given me valuable advice on these chapters and has talked to me about his work with Maxse; Lieutenant-General Sir Alexander Galloway has given me incisive and helpful criticism on what I have written on the Desert campaigns. Captain Sir Basil Liddell Hart has spared no pains to help over the Great War and has generously made available many of the papers in his library; his recollections of association with Maxse have been most valuable to me. This must not be taken to imply that he agrees with all I have written. I am particularly indebted to Mr. D. W. King, Chief Librarian of the Old War Office Library, now Ministry of Defence. Not only has he been invaluable in directing me to published works, but his memory and experience have guided me, especially in the strategic chapters. Miss Rose Coombs, Librarian at the Imperial War Museum, has spared no trouble in facilitating my search for unpublished documents. Mr. D. Chalk at the South-Western District Library, Taunton, has also been most helpful in getting me books.

I have received great encouragement and help from the Director and Lecturers of the Extra-mural Department, Southampton University, especially Dr. Edgar Feuchtwanger, Mr. Eversley Belfield, and Mr. Geoffrey Matthews. Mr. David Braddock has been most helpful in drawing the maps. Mrs. Wharmby helped in typing some of the early drafts.

Brigadier C. N. Barclay read my original synopsis and gave valuable assistance in planning the scope of the work. Major-General H. Essame and Lieutenant-Colonel Stuart Cameron have helped in reading a number of the draft chapters. My elder brother Mr. Guy Sixsmith, my wife and my two sons, and my godson Mr. Michael Hughes, have between them read all the manuscripts and have been assiduous and unmerciful in their criticism of my grammar and punctuation.

Field-Marshal Sir Claude Auchinleck, General Sir Ronald Adam, Major-General L. A. Hawes, Major-General B. P. Hughes, Major-General Sir Edward Spears, Brigadier A. C. L. Stanley-Clarke, Brigadier Sir John Dunlop, Sir Philip Magnus-Allcroft, Mr. Cyril Falls, Mr. Robert Rhodes James, Professor

Introduction

Norman Gibbs and Professor Michael Howard, have all been helpful in answering my questions. I am grateful to Major John Maxse for putting me in touch with his father's papers.

I owe a special debt to Mr. Herbert Van Thal, my agent, who in the early stages gave me valuable advice on the form the book should take. Also to Brigadier Peter Young for putting me in touch with Mr. Van Thal.

Finally I should like to pay tribute to Miss Margaret Fouracre who has typed the whole of the final text quickly, accurately and with no fuss.

A full list of the sources quoted in the text appears in the Appendix. I am grateful to the authors and publishers for the privilege of using their works. The photographs which appear in this volume are reproduced by courtesy of the Photographic Library of the Imperial War Museum, to whom I extend my thanks.

Note on Names and Ranks

Names of army officers are given without ranks unless the rank is significant or not obvious from the context. The final rank reached by the officers concerned is given in the index.

Chapter 1

THE SOUTH AFRICAN WAR

To most people Waterloo stands as the zenith of British military ability and the Crimean War as the nadir. Wellington must take his share of the blame for the rusting away of the army and for the too rigid adherence to the Wellington tradition which was at the root of much of the opposition to progress. The Duke of Cambridge, generally regarded as the arch-reactionary in military matters, saw the need, when he was a slim energetic young major-general, for change and development and did much to lay the foundations of military education and a sound staff system. But it was Wolseley and Roberts in their very different ways and often in unnecessary rivalry who brought back some inspiration and genius to British military affairs. Theirs was an era of small wars and as a conflict with the Boer Republics became inevitable it was seen as the occasion for just another colonial expedition. It was, however, to test to the full the army organized and trained by Wolseley and to require all Roberts's skill in the field. It is not only for this reason that the South African War is important, but because the war gave the British their first experience of the effect of modern rifle fire, shaped the organization of their future expeditionary force, and coloured the military thinking of the whole generation that was to rise to high command in the Great War.

In order to understand the difficulties involved in the despatch of a force to South Africa it is necessary to know the steps taken by Wolseley to form an expeditionary force when he was Adjutant-General and continued by him when he became Commander-in-Chief in 1895. Major-General Brackenbury had been given the task of preparing 'a scheme for the mobilization of two army corps and the necessary lines of communication troops for war outside Great Britain and Ire'and.' He had formed a mobilization department in the War Office when his work was interrupted by an invasion scare which turned the government's mind to home defence. Heavy naval building programmes in France and Russia had undermined the country's confidence in the ability of the Royal Navy to protect British shores. Naval measures were taken, including the construction of the 'Resolution' class battleship, but so far from the army being allowed to organize an expeditionary force it was directed to take steps for the defence of London. A ring of forts was constructed from Epping to Tilbury and continued along the line of the North Downs. These forts were little more than lightly fortified assembly points to be manned by the Volunteers while the regular army was left free for manoeuvre so as to bring the invading force to battle. The resulting policy was laid down in a confidential memorandum addressed to Wolseley in 1888 by Stanhope, the Secretary of State for War. This memorandum had so binding an effect on all attempts to organize an expeditionary force right up to the conflict with the Boer Republics that its terms are quoted in full. The objects for which the army was maintained were stated to be[1]:

(a) The effective support of the civil power in all parts of the United Kingdom.
(b) To find the number of men for India which had been fixed with the government of India.
(c) To find garrisons for all our fortresses and coaling stations at home and abroad . . . and to maintain these garrisons at all times at the strength fixed for a peace, or war footing.
(d) After providing for these requirements to be able to mobilize rapidly for home defence two army corps of regular troops and one composed partly of regulars and partly of Militia, and to organize the Auxiliary Forces not allotted to Army Corps or garrisons for the defence of London and the defensible positions in advance, and for the defence of mercantile ports.
(e) Subject to the foregoing considerations and to their financial obligations, to aim at being able in case of necessity, to send abroad two complete Army Corps, with Cavalry Division and Line of Communications. But it will be distinctly understood that the probability of the employment of an Army Corps in the field in any European War is

[1] Report of Elgin Commission, Cd. 1789 (H.M.S.O., 1903), p. 31.

sufficiently improbable to make it the primary duty of the military authorities to organize our forces efficiently for the defence of the country.

Thus it can be seen that the expeditionary force was regarded only as the residuary legatee. It was moreover made abundantly clear that if possible there was to be no residue. It is little wonder that Wolseley found this document an obstacle to all his efforts to organize an efficient expeditionary force.

At that time the strength of the regular army at home was approximately 125,000 with about the same number abroad including India. An army corps consisted of three divisions plus a cavalry regiment, artillery, engineers, and administrative units. Including the army reserve, Yeomanry, Militia and Volunteers, the strength of the army at home was about 630,000, but out of this number only two army corps and a cavalry division, 70,000 men in all, were available for despatch overseas for the reinforcement of any part of the Empire that might be attacked, or for offensive action. Moreover, as the Stanhope Memorandum made abundantly clear, both these army corps were an integral part of the field army allotted to home defence.

In judging the size of the expeditionary force it is well to remember that the largest army that had been engaged since the Crimean war was Wolseley's force at Tel-el-Kebir, rather less than one corps and one cavalry division. Kitchener's force at Omdurman consisted of two British infantry brigades and four Egyptian and Sudanese brigades, while Roberts's force at Peiwar Kotal had been even smaller. Wolseley's exchange of minutes[2] with Lord Lansdowne, Secretary of State for War from 1895 to 1900, is ample testimony that he had a very clear idea of the steps that were necessary to ensure that the army was fit for its three tasks: home defence, provision of Imperial garrisons, and the ability to put into the field an expeditionary force fit for war. In the first of these minutes Wolseley reproduced in full a minute he had written as Adjutant-General in 1888[3] demanding ' the provision of a magazine rifle for the infantry, of breech-loading guns of a modern type for the field artillery and various other most

[2] Reproduced *Ibid*. Appx D to Report.
[3] *Ibid*. p. 217.

essential requirements in the way of machine-guns, quick-firing guns, etc . . .' and recommending an increase in the establishment of the army. One sentence in Lansdowne's reply may be taken as typical of the attitude of the government of the day— perhaps of the government of any day. It read:

> I cannot lay before my colleagues any scheme involving a serious addition to our military expenditure, nor could any government venture to recommend such a scheme to parliament, until the necessity for the whole of the proposed expenditure has been amply demonstrated, and it does not seem to me that such demonstration is as yet forthcoming in regard to some at all events of the Commander-in-Chief's proposals.[4]

By June 1899 war with the Transvaal became so probable that a decision was taken that if necessary an army corps and a cavalry division should be despatched to the Cape to occupy the Transvaal. On Wolseley's advice Buller, then commanding at Aldershot, was selected to command. Wolseley pressed that these troops should immediately be mobilized in all respects except the calling out of the Army Reservists, and should concentrate at Aldershot. Such a measure would be a useful warning to the Transvaal. At the same time the necessary shipping should be taken up, establishment of transport in South Africa doubled, preliminary arrangements made for the purchase of mules and the accumulation of medical supplies and transport, and other such administrative arrangements made. None of these steps were taken. Wolseley also directed the attention of the government to the necessity of making up their minds about the attitude to be taken towards the Orange Free State.

The four months before the outbreak of war throw an interesting light on the relationship between soldiers and statesmen at that time. There were two vital questions to be decided; one, the steps necessary to protect Cape Colony and Natal if the Transvaal took the initiative; the other, the line of advance into the Transvaal if invasion became necessary. The only line of advance into the Transvaal without passing through the

[4] *Ibid.* p. 229.

Orange Free State was the route over the Drakensberg and Laing's Nek and astride the railway from Ladysmith to Pretoria. This was the shortest route and was relatively easy once the frontier passes were cleared, but it left the British lines of communication open to a flank attack from the Orange Free State. The direct route through Bloemfontein was longer but easier and could be maintained by the railways from Cape Town, East London, and Port Elizabeth. A decision to use this route depended on deciding at the outset to treat the Orange Free State as hostile. At first Lansdowne refused even to discuss this possibility. Buller was rightly insistent that a decision must be taken before any of his corps landed in South Africa. On 24 September Buller, strongly supported by Wolseley, wrote to Lansdowne urging that ' as soon as Her Majesty's Government decide upon an expedition they should force the Free State to declare for one side or the other. If they declare for the other side, our route to Pretoria should be via Bloemfontein; if they declare neutrality, they should be forced to give sureties that they preserve their neutrality. Failing to do this they should be treated as hostile.'[5] As a result Lansdowne put the question to the Cabinet and it was agreed to plan on the use of the Bloemfontein route. In the event both the Transvaal and the Orange Free State began hostilities on the same date, 11 October 1899, before Buller had left England.

Wolseley never had any doubts that the Aldershot force would be sufficient to complete the defeat of the Transvaal. He was however from the first doubtful whether the forces then in South Africa were sufficient to hold the situation before the arrival of the corps. The danger was not so much in Cape Colony—except for insurgency from Boar sympathizers—as in Natal. In June 1899 the garrison of Cape Colony was two and a half infantry battalions plus the coast defence troops at Cape Town. In Natal there were two regiments of cavalry, a Brigade of artillery (three field batteries and a mountain battery), and three infantry battalions. Sir William Butler commanding in South Africa had been warned in December 1898 that, in view of the deteriorating situation, he must prepare the defence measures which would be necessary before the arrival of an Expeditionary Force from England. Butler who had only just

[5] *Ibid.* vol. II, p. 505.

arrived, and in the absence of Sir Alfred Milner, was acting as High Commissioner, did not take the danger at all seriously; broadly speaking he regarded the Boers as having justifiable grievances which could easily be put right by the British Government. From a military point of view he regarded his duties as making the best use of the troops at his disposal. He proposed to do this by dispersing the garrison in Natal to protect approaches and communications, to guard against raids, and if necessary to withdraw and concentrate at Estcourt, south of the River Tugela, in order to cover the approach to Durban. In July 1899 there was a meeting of Lansdowne, Wolseley, and Buller in which Buller was asked whether the position in Natal and Cape Colony could be considered safe if an ultimatum were sent to the Transvaal. Buller replied that he had complete faith in Butler and others, and that so long as they did not say there was any danger he did not see any necessity to send out any troops in advance of the army corps.[6]

As a result of differences with Milner, Butler was relieved of his appointment in August 1899, but by this time matters were being taken out of the hands of the local commanders. There was a good deal of correspondence between Milner and Lansdowne only part of which was seen by the Commander-in-Chief. Lansdowne was anxious that Wolseley's eagerness to get a force out to South Africa should not prejudice the negotiations with the Transvaal. On 18 August, however, Wolseley recommended that in view of the deteriorating situation a division and some additional cavalry and artillery should be sent at once to Natal. Lansdowne, replying from Ireland, expressed regret that this recommendation had not been put forward before the Cabinet separated (presumably for the start of the grouse shooting season on the 12th) and pointed out that the situation in South Africa had much improved. He did not think reinforcements were necessary at that time but if they became so he suggested that 10,000 men should be sent from India where they were in readiness. Wolseley did not agree that the situation was any easier and said Kruger was clearly making preparations for war; he added: ' I still believe, perhaps foolishly, that a display of force would be the quickest and surest way to ensure peace.' He did not believe it would be quicker to send troops from India

[6] *Ibid.* Appx D to Report, p. 264.

as shipping could be obtained more quickly and cheaply in Britain. Moreover, he preferred the battalions from England with their young soldiers to the more experienced but less well trained and less fit men from India. Lansdowne was still against sending reinforcements until it was clear we could not come to terms with the Transvaal. In consequence, on 5 September Wolseley sent on to the Prime Minister, Lord Salisbury, a letter from Buller stressing that diplomacy and military policy must go hand in glove and that it was no good sending an ultimatum until they had decided what action to take if the ultimatum were rejected and were in a position to take it. The military appreciation was that the Boers could act in three weeks while the British required five weeks to get 5,000 men from India, eleven weeks to get 10,000 men from the United Kingdom, and sixteen to twenty weeks for the earmarked army corps. In addition transport for operations in South Africa would require special organization which would take time.

On the same day that he addressed the Prime Minister, Wolseley minuted Lansdowne as follows:

> The first intimation I have had that our negotiations with the Transvaal have reached an acute stage has come to me from Buller.... Can we not stave off actual hostilities for five or six weeks to enable us to collect in Natal the military force I have all along recommended should be sent there? We have committed one of the very greatest blunders in war, namely, we have given our enemy the initiative.... I have not yet been told if the Indian contingent has been ordered to Natal but assuming this has been done, it could not, I believe, be in line at Ladysmith before 18 October.[7]

He recommended that a battalion from Cape Town and another from Malta should be sent to Natal at once, both to be replaced from home. Sir George White, Quartermaster-General at the War Office, should be sent at once to command in Natal, becoming second-in-command to Buller when he arrived, and Major-General Hunter should be sent at once from India to

[7] *Ibid.* pp. 268-9.

be Chief of Staff to Buller. The reinforcements from India were already on their way and Lansdowne accepted Wolseley's recommendations. At the same time Lansdowne asked Wolseley whether the army corps, if it were sent from Aldershot, could be reduced in view of these reinforcements amounting to about a division. Wolseley strongly and successfully resisted this reduction despite strong opposition in the Cabinet; only a few days before war broke out Lansdowne had to write a strong memorandum to the Cabinet defending his decision.

On 11 October all doubts were set aside by both the Transvaal and the Orange Free State as they commenced moving to the invasion of Natal. White had arrived in Natal on 7 October; the only instructions he received were that his command was independent of Cape Colony. Buller's corps was not mobilized until 7 October, and he himself sailed in advance of it on 14 October. By 12 October, however, all reinforcements sent in advance of the main force had arrived in Natal so that White had about a division plus a cavalry brigade. He was immediately faced with the problem of how far forward to engage the enemy. The pre-war plans had held half the force forward at Glencoe with the remainder at Ladysmith and Estcourt, covering Pietermaritzburg and the route to Durban. Ladysmith was provisioned as a base from which forward operations could be maintained in the Transvaal, or through Van Reenen's Pass into the Free State. Ladysmith was valuable as a forward base because it was the railway junction and the main town in Northern Natal, and because of the water supply. It was thus of some strategic importance to both sides, but as a defensive position it left much to be desired because, although surrounded by a ring of low hills which gave it a local perimeter of some strength, it was in low ground extensively overlooked. It could also be easily by-passed to reach the River Tugela at Colenso. White wished to withdraw from Glencoe and concentrate on the defence of Ladysmith. The Governor considered such action would be disastrous since it would cause the Dutch in Natal and possibly in Cape Colony to throw in their lot with the Boers and would lead to the disaffection of the Zulus and possibly other Africans; it would also mean the abandonment of the rich Dundee coalfield. In the event White left the garrison at Glencoe where they fought two successful but costly actions at Talana and

Elanslaagte on 20 and 21 October after which they retired on Ladysmith. White was then confident of being able to take offensive operations against the Boers but later the failure of a somewhat complicated attack was regarded by him as a major defeat and by 30 October he had allowed his whole force to be besieged in Ladysmith.

After the war, in criticism of this action, Roberts considered that White should have met the Boer forces while they were still separated north and west of Ladysmith and defeated them in detail. In justice to White it must be remembered that he had some 13,000 men about Ladysmith, most of whom had only just arrived in the country, while the force from the Transvaal numbered at least 15,000 and that from the Orange Free State about 10,000. White's difficulties were increased because his force was surrounded by spies. Wolseley's criticism was that White should have held the strong Biggarsberg position north of Ladysmith and when he could no longer do that he should have withdrawn to the Tugela. Faced with these apparently conflicting opinions White was able to say: ' When two such high authorities take views so exactly opposite of what the right course was, it may, at all events, be allowed that strategy is not a very positive or exact science, the study of which leads to uncontentious conclusions.'[8] White can hardly escape with this specious argument because Wolseley and Roberts were both really saying the same thing: that Ladysmith was a useful base for operations in the north but not an essential defensive bastion. It is true that Roberts in his evidence, with the kindly tolerance which characterized all his evidence, defended White's efforts to try to hold a base which had been made before his arrival, but nothing he said could be taken to support White's action in allowing himself to be shut up in Ladysmith. This is an example of a base well chosen for an offensive which, when the enemy is able to take the initiative, has a cramping effect on the defensive. We are to see almost exactly the same thing in 1942 at Tobruk.

By 31 October when Buller arrived in Cape Town not only had the Ladysmith battle been lost, but Mafeking was besieged and also Kimberley, where the large civil population included Cecil Rhodes and the diamond magnates. Orange River Bridge

[8] *Ibid.* vol. 11, Evidence, p. 146.

on the Kimberley railway was still in British hands but the railway crossings into Cape Colony from the Free State had been captured intact by the Boers, and Milner was apprehensive of a rising in Cape Colony. On the voyage Buller had not discussed his plans with his staff but before leaving he had had many consultations with Wolseley, who believed that the best course to take was to advance astride the central railway on Bloemfontein and then on to Pretoria. Wolseley believed that is what Buller would do, although he was careful not to fetter in any way the initiative of the chosen commander. In retrospect Wolseley still believed this would have been the right course. A rapid advance to the heart of the Transvaal would have forced the Boers to concentrate for its defence and the capture of Bloemfontein would have created the situation in which Kimberley and Ladysmith could not have remained invested. Buller's appreciation of the situation after White's defeat was very different. His force would not be fully concentrated and ready to move on Bloemfontein until 22 December. In Natal there was fear even for the safety of Durban, and he felt he could not entirely neglect Kimberley. He considered, moreover, that his true objective was the main Boer force and this was now concentrated in Natal. On 10 November, therefore, he reluctantly decided he must divide his force. Methuen with 1st Division was ordered to march on Kimberley, throw in provisions, take out non-combatants, and return to Orange River Bridge. Gatacre, commander of 3rd Division, which had not yet arrived, was to hold the approaches along the central railway to East London with a scratch force of only one battalion, later increased to three and a half battalions. French with the Cavalry Division was to remain in the centre near Naauwpoort. There he was to train and to take every opportunity of harrowing the enemy. He was instructed that if possible he was to drive them out of Colesberg and over the Orange River but that he was not to endanger his force. Buller himself with 2nd Division went to Natal to restore the situation there. By this division of his forces Buller straight away surrendered the initiative and none of his columns achieved its aim. On 10 December Gatacre was defeated trying to clear Stormberg and the next day Methuen was defeated at Magersfontein. Buller was at this time preparing a flank attack against the forces besieging Ladysmith but these reverses so disconcerted him that he no longer dared to take the risk of cutting loose from his communications. He therefore decided on a direct advance on Ladysmith and was defeated at the battle of Colenso. Buller then seems to have lost his head; he recommended to the War Office that Ladysmith should be abandoned, and suggested to White that he should shoot off all his ammunition, burn his cyphers, and then surrender—a suggestion which White immediately repudiated. Buller's reasoning was that it would be better to lose Ladysmith than to take risks which might throw the whole of Natal open to the Boers. Wolseley, strange to relate, took no initiative for the supersession of Buller in command, while even more surprisingly Lansdowne without consulting Wolseley discussed with the Prime Minister the appointment of Roberts to go out as Commander-in-Chief. Most surprising of all is that when Roberts did go out, with Kitchener as his Chief of Staff, Buller remained in command in Natal.

The subsequent operations in South Africa are of less strategic interest, although they abound with tactical lessons. Buller's force had been considered adequate for the war in South Africa but Wolseley was prudent enough to ensure that as soon as it had sailed a further division be mobilized, and in turn all the divisions of the second corps were mobilized and despatched. Thus Roberts had at his disposal seven divisions (including White's force) and a cavalry division. In addition a large number of Militia and specially organized Volunteer and Imperial units took part in operations or on the lines of communication. In all a force of more than 440,000 (of whom 256,000 were regulars) was used to defeat an enemy that never exceeded 65,000.

Roberts decided that his best course was to move with his main force on Kimberley. This allowed him to use the only bridge over the Orange River in his hands and after the relief of Kimberley would enable him to strike across country to Bloemfontein and on to the main route into the Transvaal. He was confident that his advance would regain for him the initiative and would enable Buller to link up with White at Ladysmith, and that Baden-Powell would be able to hold out in Mafeking. Roberts was able to carry out his operations as planned, though not without difficulties when he cut across from the Kimberley railway to Bloemfontein. Nevertheless, as he expected, this move led at last to the relief of Ladysmith by Buller and a few days later, on 13 March, to the capture of Bloemfontein. After a pause to reorganize his supply and communications, Roberts moved on to Pretoria, which was occupied on 5 June.

Roberts's performance in command in the operations up to Pretoria can be belittled on the grounds that he always had superior forces, and that he never achieved a major victory. Although the cavalry under French was nearly always used on the flank to cut off the enemy, decisive action never resulted. After the relief of Kimberley, French did succeed in heading off Cronje's force, which was trying to cross the Modder River. During the resulting battle of Paardeberg Roberts was unfortunately ill and Kitchener commanded. Kitchener failed in his attack. This was his first battle against a white enemy and his only failure in battle. Roberts decided not to renew the battle but to invest Cronje, who surrendered nine days later. Paardeberg was the only major battle in the war, but ten days after Cronje surrendered the real opportunity to end the war was missed. The cavalry under French was ordered to make a seventeen mile march round the Boers' southern flank to cut them off while two divisions made the main attack. The plan was boldly conceived but marred by French's slowness and hesitation in action. The official history attributes this failure to the horses' lack of condition, but it seems more likely that French was below his best. It is known he was angry over an unjust reproof he and his staff had received over the rations for his horses. This was a foretaste of the uncertainty of temperament which French was to show in 1914. Included in the Boer force which the British allowed to slip through their fingers at Poplar Grove were Kruger, Steyn, and de Wet; it is unlikely that the Boers could have continued the war without them.

Wolseley, pressed by the Elgin Commission to comment on Roberts's strategy considered that he was wrong to relieve Kimberley first and then strike across the Modder River to the central railway. It is true that many of Roberts's difficulties occurred in the operations on this flank march across country and that after the capture of Bloemfontein there was a six weeks' delay before transport and supply could be organized for an advance on Pretoria. This gave de Wet just the opportunity he needed to reorganize his forces and to play havoc in the area between Bloemfontein railway and the Basutoland border. Roberts never got the better of de Wet and his commandos, but to his everlasting credit he was not deflected from his main purpose and as soon as he was administratively ready he continued the advance on Pretoria and thus retained the initiative. Wolseley was almost certainly right in the advice he gave Buller before he left England but there is little justification for his view that the same strategy would have been right for Roberts. When Roberts took over command the Orange River Bridge was in British hands but all the crossings for a direct advance on Bloemfontein were in the enemy's possession. A direct advance would have involved the same hazards as had led Buller to disaster at Colenso. There is no doubt that Roberts was one of the first to see the changes which modern weapons had brought to warfare, and that he was more at home in manoeuvre than in a head-on collision. He was not the man for a battle of attrition such as Haig was to direct in the next war. Roberts's skill lay in the fact that, arriving late to command at a moment when everything was going wrong, he gained and retained the initiative without incurring more casualties than were inevitable.

After the fall of Pretoria Roberts was not the only one to consider the war at an end and he was recalled to succeed Wolseley as Commander-in-Chief. But the war dragged on for more than eighteen months with Kitchener in command striving to defeat the Boer guerrilla activities by a system of blockhouses and mobile patrols. The country was divided into areas separated by wire fences and cleared by systematic drives. Kitchener judged the success of his operations by the weekly bag of prisoners. Although the people and the country are so different the military technique is what we are to see again in Malaya from 1952 to 1954.

The tactical lessons of the war were those that men like Wolseley—and certainly Henderson—had begun to learn from the American Civil War. The invention of smokeless powder had accentuated the effectiveness of riflemen skilfully hidden and entrenched, but the importance of mobility and the necessity for initiative in junior leaders was even more strongly brought out. No longer could battles be won simply by massing superior numbers of determined troops at the right place. Surprise, tactical skill in the use of ground, and the opportunism which cunning leadership could develop were necessary for victory in the face of modern weapons. And yet, whatever the skill employed, nothing would avail unless at the last commanders were determined to press the attack and men were disciplined to obey in spite of casualties. It is this paradox, this balance between the wits and the will of the opposing commanders, that makes war so difficult and its practice so different from arm-chair strategy. Henderson had seen the two sides of the problem. His insistence on the importance of mystifying and misleading the enemy are well known; it is not so often remembered that he also wrote: 'Well has it been said that the company leader who regardless of losses carries out the task assigned to him, is a better servant than the company leader who manoeuvres.'[9] The balancing of the need for manoeuvre and determined attack, and of attack and defence is the one respect in which the Boers showed themselves an inferior enemy. They were most skilful in the use of cover, of entrenchments, and of weapons in defence, but when they had defeated the attack they never on any occasion counter-attacked. Had they done so at Magersfontein and on numerous occasions against White and against Buller these campaigns would have been even more disastrous. The Boers' failure to use the offensive in any form except as raids shows no more than the weakness of the irregular against the trained army, but it may have suggested wrongly to some of our commanders that the use of cover and entrenchments saps the offensive spirit.

A similar difference in interpreting the lessons of the war appears in the use of cavalry. All admitted that the war showed the paramount importance of mobility, and all agreed that the cavalry must be armed with the rifle or carbine. But despite the fact that there was only one example of a successful cavalry charge there were those who contended that cavalry could not be imbued with the offensive spirit unless trained and ready to charge with sword or lance. Before the Elgin Commission

[9] Henderson, *The Science of War* (Longmans Green, 1908), p. 147.

many diverse opinions were given. French and his chief of staff Haig wanted the sword preserved. Haig foresaw that the use of effective firearms gave cavalry a larger sphere and an independent strategic role.[10] Both were sure that cavalry and mounted infantry should be kept separate and that it would be a mistake to withdraw the sword from the cavalry. Opportunities for the use of the ' arme blanche ' would be few but of paramount importance in morale. Haig said: ' To take away from cavalry the power of assuming the offensive by mounted action by depriving it of the " arme blanche " is to withhold from it a very considerable advantage without a compensating gain.' Ian Hamilton on the other hand said: ' Compared to a modern rifle the sword can only be regarded as a mediaeval toy.'[11] Roberts wanted to abolish the lance but was willing to let the cavalry keep the sword and train for mounted action. However, he said: ' From my own experience and what I have been able to read up, and understand from wars, certainly from Napoleon's time, there has very seldom been much done by shock tactics, and now that we have long ranging rifles it is more than ever necessary that the cavalry should be taught to act dismountedly.'[12]

The development of tactics showed that the officer had a completely new role to play. It was no longer sufficient for him to bring his men to the point of battle and then lead them blindly against the enemy. Both staff officers and regimental officers must be professionals in every sense of the word. Wolseley had long preached this, and now Roberts, in summing up the performance of the army in the war, said:

> A certain proportion of failure in war is inevitable; but among the subordinate regimental officers in South Africa it was extraordinarily small . . . I should be the last to say that there is no room for improvement. The first point is that officers should take their profession more seriously than has hitherto generally been the case, and that they should be able to instruct their men in every detail of their duty. The second point is a wide knowledge of war. . . . Education in the army stopped short at the drill books; history was a closed volume,

[10] Report of Elgin Commission, vol. II, Evidence, p. 403.
[11] *Ibid.* p. 110.
[12] *Ibid.* p. 66.

except to those who opened it for themselves... The proportion of failures among Commanding Officers and Brigadiers was considerably higher than in the junior ranks... As men get older they are often less inclined to accept responsibility, and they lose their power of decision. On the staff those who had received previous training, either in active service or at the Staff College, generally did well; but the absence of a definite system of staff duties, leading sometimes to an overlapping of responsibilities, sometimes to waste of time, and sometimes to a neglect of indispensable precautions, was undoubtedly prejudicial to a smooth running of the military machine... Staff Officers cannot be improvised... I am decidedly of the opinion that we cannot have a first rate Army unless we have a first rate staff, well educated, constantly practised at manoeuvres, and with wide experience. Brains are even more important in war than numbers; and in an army which may contain a large proportion of men who are not soldiers by profession, trained leaders are especially important.[13]

On the wider issues the lessons were largely those of relationship between Government and its military advisers, and between both of them and the commanders in the field. The war came when Wolseley was past his best and brought for him the great bitterness of seeing the conduct of the war and the fruits of victory going to a man older than himself; going moreover to a man he had always considered, perhaps not without reason, to be militarily his inferior. But in at least two directions the efforts of Wolseley to reorganize the army and increase its efficiency had borne valuable fruit. His Mobilization Department and his Intelligence Department had done all that could be expected of them. The mobilization of the First Army Corps and the subsequent divisions went without a hitch; the later contingents for an expansion not hitherto contemplated were provided by hand to mouth methods using the Militia and the Volunteers, and overcoming all sorts of difficulties for the provision of material. The main weakness on this side was in the provision of transport. Things would have been much better if Wolseley's warnings had been heeded. Detailed as the

[13] *Ibid.* vol. I, Evidence, p. 441.

mobilization arrangements were they could not go beyond the use of the Regular Army Reserve. For the lines of communication it was not only necessary to improvise arrangements for horses and wagons but also to embody drivers on a scale not hitherto contemplated. The difficulty was only surmounted by using civilian drivers in the field, an expedient which would not have been possible but for the fact that the enemy did not wear uniform.

It may be asked how the Intelligence Department can be upheld. In his evidence before the Elgin Commission, Kitchener made a great point of the fact that the Intelligence Department had underestimated the force which the Boers could put into the field against the British. He was given time to produce a statement of the numbers they actually did field and it coincided almost exactly with the estimates by the Intelligence Department made in 1896 and 1899.[14] The steps which Wolseley had taken to appraise the government of the true situation, and the steps necessary to maintain the British position in South Africa have already emerged in this chapter. Where Wolseley did fail was in the selection of his generals. Not only Buller and White but also almost all the Divisional Commanders failed. Warren, Clery, and Gatacre failed miserably and Methuen can certainly not be cleared of all criticism. Wolseley wrote of Buller in February 1900: 'Thoroughly disappointed in him for he has not shown any of the characteristics I had attributed to him, no military genius, no firmness, not even the obstinacy which I thought he possessed when I discovered him. He seems dazed and dumbfounded.'[15]

To read Buller's minutes and letters before he left for South Africa is to see a man apparently master of the situation, but from the moment of his arrival he showed himself incapable of taking effective action. After the reverses suffered by Methuen and Gatacre he lost his head completely. He was always telegraphing for advice—to Lansdowne, to White, to Roberts, but oddly not to Wolseley. When he got their advice he could always think of many reasons against taking it. The amazing thing is that despite all his defeats and mistakes Buller never seems to have lost the confidence of the rank and file of the

[14] *Ibid.* p. 13.
[15] Letter to Lady Wolseley, R.U.S.I.

army or of the nation as a whole. Possibly the reason for this is that Buller was always most careful of his men's lives. Like Roberts he hated the direct assault, but he lacked Roberts's skill in directing his manoeuvres so as always to maintain the initiative. White was a protégé of Roberts not of Wolseley. In India Roberts had regarded him as his most able and daring subordinate, but the choice was Wolseley's. Both Buller and White were holders of the Victoria Cross. It must be remembered that although mobilization arrangements existed for two army corps, units were not organized in divisions in peace time. Commanders for them and officers for higher appointments had therefore to be collected from all over the army. Wolseley, who had made such a point of selection by merit rather than by seniority, can be accused not of choosing only his favourites of the Wolseley Ring but of looking too much to his own generation. At the beginning of the war Buller was within a month of his sixtieth birthday and White was sixty-three. When Wolseley had sailed for the Ashanti Expedition he and all his staff had been under forty. The opposition that Wolseley had met from the Duke of Cambridge and others had left a lot of dead wood in the army, but if only Wolseley had realized that there were officers no younger than he had been in the Ashanti, fit and eager for executive command, how much greater would have been his service to his country and the army!

Chapter 2

ARMY REORGANIZATION AND THE CREATION OF THE EXPEDITIONARY FORCE

After the South African War there was a great upsurge of interest in the army and in Imperial Defence. There were several reasons for this. The war had not been a war of nations in the sense that was to come but it had proved beyond the capabilities of the regular army alone. Citizen armies had come into their own; special improvization had been necessary to enable the Militia and the Volunteers to take part in the war overseas, and specially raised units from the Empire had come to the assistance of the Mother Country. Most significant of all, the effort of dealing with two small South African republics had left the British defenceless at home except for their navy. The nations of Europe had not concealed their delight in their difficulties; above all, Germany, where there was great interest in British performance and in the lessons of the war, had begun to show herself determined on supremacy in Europe and further afield. Thus, at the time when the British had had a sharp military lesson, their diplomacy saw the end of 'Splendid Isolation' and, for the first time for a hundred years, the fear of direct involvement in a major continental war.

The reference in the last chapter to the evidence given by soldiers before the Elgin Commission shows the extent to which thought was being given to the lessons of the war. But the Commission had a wider purpose than to study tactical lessons. It was directed to enquire into 'the military preparations for the war in South Africa, and into the supply of men, ammunition, equipment, and transport by sea and land in connection with the campaign, and into the military operations up to the occupation of Pretoria.' The Commission interpreted these terms of reference by assuming that the object of the enquiry was to discover inefficiency or defects in the civil control and the administration of the army and their causes. Evidence was taken from Lansdowne, from Brodrick, who succeeded him as

Secretary of State for War, and from all the heads of departments in the War Office. While no attempt was made to compile a history of the war the Commission did 'obtain the evidence of the leading actors in the war arranged, as far as possible, in chronological sequence'. Thus the fact that statements were taken from White, Buller, and Roberts, and that in his evidence Wolseley gave some comments on the conduct of operations, made the report a most valuable commentary on the conduct of the war. It also fulfilled its primary purpose as an examination of the working of the national machinery for control of the army and for making war. The Report, including full minutes of evidence and the appendices was laid before parliament in 1903.[1]

Among the members of the Elgin Commission was Viscount Esher. With Esher army reform was an obsession; he had been private secretary to Hartington from 1878 to 1885, covering the time when Hartington had been Secretary of State for War and Chairman of the Hartington Committee. Esher refused Balfour's offer that he should become Secretary of State for War, but he did accept appointment as chairman of the Prime Minister's committee on War Office Reform, independent of the Secretary of State for War, Arnold-Forster. Two far reaching reforms came directly from the Esher Committee. The first the abolition of the office of Commander-in-Chief and the institution of an Army Council; the second the creation of the Committee of Imperial Defence. Both reforms were to some extent the fruits of the Hartington Committee, which had wished to separate the command function from the War Office. It had proposed therefore to abolish the office of Commander-in-Chief and to set up a War Office Council consisting of a Chief of Staff and heads of Departments each responsible to the Secretary of State. No Prime Minister had felt strong enough to recommend this change during Queen Victoria's lifetime, so when at last the Duke of Cambridge could be persuaded to give up the office Wolseley was made Commander-in-Chief with much reduced powers. The South African War had clearly shown the consequences of this equivocal policy with its failure to decide and define the source from which the Secretary of State was to get his official advice. Now Esher persuaded the

[1] Elgin Commission (4 vols. Cd. 1789-1792).

Reorganization and the Creation of Expeditionary Force

Prime Minister to face, for the first time, the problem of the relationship between policy and military preparations. The setting up of the Army Council, collectively responsible for all affairs in the army, was the essential first step. This achieved, Esher sought to create the machinery for the proper consideration of plans and preparations for war. In the first place the Committee of Imperial Defence, with a naval and military secretariat, was instituted so that defence matters could be dealt with as one with Cabinet, Army, and Navy facing the problem together. Secondly a General Staff was set up for the army. Colonel Ellison, who was secretary of the Esher Committee, had worked out a staff plan for Roberts when he was Commander-in-Chief in 1901. This followed the old Wellington tradition in which the Quarter-Master-General was the executive and operations staff officer. Under Wolseley the Adjutant-General had become the principal staff officer, leaving the Quarter-Master-General with little but quartering and movements. Now the General Staff embraced operations, war organization, and training; the Adjutant-General remained responsible for discipline, ceremonial, and questions of personnel; the Quarter-Master-General had the added responsibility of supplies and transport; the fourth military member, the Master-General of the Ordnance was responsible for the provision of armaments and equipment. The organization set on foot by Esher has remained unchanged except in detail to the present day; indeed it may be said that the present integration of the forces has its seeds in the Esher Reforms—from the Committee of Imperial Defence through the Chiefs of Staff Committee and the Joint Planning Staff to the Ministry of Defence.

The Balfour Government which had so firmly grasped the question of War Office reform was not so successful in deciding the shape of reserve forces necessary to meet the changing situation in Europe. By 1904 the Entente with France was a reality, and fear of a Russian threat to India was lessened by the Russo-Japanese war. But the British were by no means committed to military involvement on the Continent. Even the staff conversations with France, which began in 1906, were undertaken without political commitment. The realization that their regular army was not by itself sufficient to back their foreign policy was combined with a failure to agree on the extent to which the nation should be involved in military service. For the first time thought was being given to the necessity for some form of conscription. Roberts, freed from responsibility by the abolition of his post as Commander-in-Chief, threw himself into the leadership of the National Service League, which had been formed in 1902. He believed that compulsory military service was essential to national survival. There were others who considered that a conscript army was unsuitable for our varied commitments overseas, and who believed that the situation could be better met by a reorganization of our volunteer reserve forces. Two successive Secretaries of State, Brodrick and Arnold-Forster, and the War Office staff saw that there was no logic in the organization of our army at home. Forces available for overseas and those for home defence were inextricably mixed; the Militia and the Volunteers both existed for home defence but were not organized into higher formations than battalions. It was easy to see that what was wanted was a Militia organized as a reserve for regular units in an expeditionary force, and Volunteers properly organized into a field force for home defence. But this required legislation and comprehensive reorganization. The Militia would have to be changed from the 'Constitutional Force', based on the idea that it is the obligation of every man to defend his country in time of need but that overseas expeditions are the business of the King's Army—the regular forces. Moreover the reshaping of the Volunteers required the abolition of a number of units and the change in role of a number of others in order to meet the cost of a field army. Both measures met with determined opposition in parliament. Opposition to changes in the Militia was strongest in the House of Lords, while the Commons resisted the reduction in the number of Volunteer units. The question was not resolved until there was a change of government, when Haldane became Secretary of State for War. Haldane, a lawyer and a philosopher with a brilliant analytical mind, approached his task from first principles. He tackled the question of the organization of an expeditionary force and of the reserve forces without preconceived ideas or prejudices. He was prepared to proceed cautiously but he quickly gained the confidence of the War Office Staff. Military members asked him at their first meeting for his ideas on reform. He replied: 'I am a young and blushing virgin just united to a bronzed warrior and it is not expected by the public that any result of the union should appear until at least nine months have passed.' A few months later he wrote: 'The dear generals are angels ... I have already made changes which might have tried them, and they gulp them down ... you never saw such a band of reformers as I am trying to hold back. If I could only get three years I could do something.'[2]

In the War Office the detailed work of the reorganization was in the hands of the Director of Staff Duties, Major-General Douglas Haig. Between Haldane and Haig there was an understanding and mutual respect which assured the proper balance and co-ordination between political and military factors. As a result was created the Expeditionary Force of six divisions and a cavalry division, supported by a Special Reserve, in place of the Militia, with the function of supplementing the Regular Army Reserve. Behind this, primarily for home defence but organized for any field role after embodiment and training, was a Territorial Force of fourteen divisions and fourteen Yeomanry brigades. It is to be observed that even Haldane, with all his political sagacity, was unable to persuade the Militia either to form part of the Territorial Force or to adopt the role for which the new Special Reserve was created. The Militia chose rather to go into abeyance, and, as the last officers and men came to the age limit or the end of their engagement, it disappeared from the Army List. The essential merit of the Haldane reforms was that now, for the first time, the British had their army serving in peace in the formations and under the commanders with which they would fight in war.

Despite the wholehearted support which Haldane received in the War Office, critics were not lacking. Major-General Henry Wilson, Commandant of the Staff College—later to be Director of Military Operations during the staff talks with the French—was fond of saying that there was no military problem to which the answer was six divisions and a cavalry division. He was also contemptuous of the Territorial Force. He believed it impossible, for example, for a part-time army to provide efficient field artillery and field engineers. Wilson's criticism

[2] Sir Frederick Maurice, *Haldane*, vol. 1 (Faber), pp. 169, 158.

was no doubt based on his knowledge that the size of the Expeditionary Force was not based on an operational assessment of what was required. It was based on the size of the army that could be created with the number of regular soldiers required in Britain for the maintenance of peace-time commitments in the Empire overseas. No doubt Haldane knew this too. If he had set out to produce a full scale continental army he would have achieved nothing. British traditional strategy was based on the sea and on alliances which would enable them to limit their participation in a continental war.

Kitchener too was critical. During all the period of the Esher and Haldane reforms he kept aloof. He had a supreme contempt for politicians and had already shown that he could work only where he was in undisputed control. When the Army Council and the General Staff were established he was the obvious choice for Chief of the General Staff, first military member of the Army Council. He refused the appointment because he did not believe in joint control; in his letter to Esher refusing he said:

> 'What I want to impress upon you is that it would take a great deal to convince me now that it was my patriotic duty to accept the post of C. of G.S. Why? Because I should fail! I think I know what I can do, as well as my limitations. I can, I believe, impress, to a certain extent, my personality on men working under me; I am vain enough to think I can lead them, but I have no silver tongue to persuade...'[3]

This refusal came just after Kitchener, as Commander-in-Chief in India, had emerged victorious from his conflict with Curzon, the Viceroy, over military membership of the Viceroy's Council. In this controversy Kitchener had shown himself irrevocably opposed to divided control or advice by committee on military matters.

Despite his great military reputation Kitchener seems to have made little impact on the training of the army and the development of military technique. It is true that as Commander-in-Chief in India he did establish a Staff College at Quetta to work on exactly the same lines as Camberley. Both Staff Colleges

[3] Esher, *Journal & Letters,* vol. II, p. 98.

Reorganization and the Creation of Expeditionary Force

were attended by both British Service and Indian Army officers, and later by officers of the Dominions, while the qualification at each had exactly the same status. But Kitchener believed in personal autocratic control; in his own words ' you rarely get two soldiers who think alike even on the fundamental military questions'![4] He seemed to have a mind above the creation of a common military doctrine for the army. Kitchener's lofty indifference to tactical thought is shown by an account which Ian Hamilton gives[5] of a visit to Mhow in which Kitchener as Commander-in-Chief was present at a field exercise of all arms. The artillery were sited in the open, engaging targets by direct fire, the cavalry were indulging in spectacular charges, and the infantry were forming up for attack widely extended as was the practice in the South African War. All this, Hamilton, who was on his way home from his duties as Observer in the Russo-Japanese war, knew to be years out of date. Hamilton prepared notes for Kitchener on the importance of trenches, wire, attacks in depth, indirect fire, and the futility of cavalry charges. Kitchener was not in the slightest degree interested. He allowed the tactics of a former day to pass unchallenged, and departed without a word of criticism for the exercise.

The greatest impact on the tactics of the day came from Roberts, whose understanding of the lessons of the South African War has already been indicated. After the war Roberts had set Henderson to rewrite the Infantry Drill Book. Henderson died before his work was completed but it was used extensively for the production of the new training manual known as *Combined Training*. This was produced in the department of 'Military Education and Training' which Roberts formed at the War Office and to which he appointed both Rawlinson and Henry Wilson. Those who read this manual[6] with the preconceived idea that the 1914-18 War was grossly mismanaged will find little in it to suggest that the fault lay in pre-war tactical thought. The importance of entrenchments is foreseen in such phrases as: ' It should generally be comparatively easy for the defender to hide his dispositions, and to effect a surprise. Entrenchments, moreover, are more

---

[4] To Lady Salisbury, 28 Oct. 1903. Hatfield MSS.
[5] Hamilton, *The Commander* (Hollis & Carter, 1955), p. 120.
[6] The extracts which follow are taken from Combined Training Pt. 1 Field Service Regulations 1905, ch. VI.

easily employed by the defence, and by their aid a position may sometimes be rendered practically impregnable.' The interdependence of the various arms in battle is stressed, and on surprise we read: ' Surprise, then, tactical as well as strategical, should always be the aim of an officer commanding in the field, whether he is attacking or defending.' In planning an attack on an enemy in position the commander is advised, whenever possible, to outflank one of the enemy's wings and ' to seize localities from which a searching and sustained fire may be developed against a weak point of the position; to strike at that point heavily, unexpectedly, and in the greatest strength possible.' But cognizance is taken of the difficulties that an entrenched position on a wide front and in great depth would offer. ' In such cases a study of the weakest points in the defence should indicate a succession of vantage points which should be attacked in methodical sequence, one advantage gained leading to the next.' Throughout this section emphasis is placed on attacking the enemy where he is most vulnerable. Once the point of attack has been selected, however, no doubt is left in the reader's mind that the attack must be pressed home despite casualties. Thus reference is made to the points selected for assault, ' and which must be carried, cost what it may.' And later: ' The critical moment has now come. The artillery and machine-guns pour in the most rapid fire, and orders are given for the final advance of the infantry. Strong reinforcements are thrown in, and as they reach the firing line, carry it with them, and pressing forward with the utmost vigour and resolution, regardless of losses, rush the position.'

The section on the action of the cavalry and other mounted troops makes no mention of the charge. Cavalry are given the tasks of protection and reconnaissance and ' to seize, and hold tactical positions, often far in advance of the slower moving infantry, and to deny their occupation to the enemy until the infantry arrives.' It is made clear that these primary tasks of cavalry may prevent their being available to take part in the main battle. If they are available they are to be used on the flanks and ' during their flanking movements cavalry should endeavour to co-operate by means of fire with the main attack ... Or, again, when the ground is favourable, it may be directed to pass right round the flank of the enemy position, and operate against his reserves and communications. The possibility of using the cavalry in sustained pursuit of a defeated enemy must always be in the mind of the commander.'

In 1909 a revised training manual was issued, this time called 'Field Service Regulations Part 1 Operations'. It propounds much the same doctrine but this time we miss some of Henderson's lucid and illuminating style, which the editors of 1905 had retained. Moreover, there is not the same emphasis on surprise although its importance is stressed in defence, in the counter-attack, and from concentrated fire of machine-guns. Much is made of the importance to the commander of the initiative: in this context, 'A successful cavalry will retain for a commander the initiative he has gained, or regain it for him if it has been lost; it will gain him strategic liberty of action, and will thereby enable him to act with certainty and impose his will upon the enemy.'[7] The dominant part to be played by the cavalry is again stressed in the section on the attack. 'As the crisis of the battle approaches, and the enemy becomes morally and physically exhausted, the chances of successful cavalry action increase. For effective intervention the concentration of as large a part of the cavalry as possible is required.' This may be considered somewhat at variance with the 1905 manual, quoted above, which foresaw primary tasks preventing immediate participation in the main battle.

This section on the attack has a more ponderous flavour than that of 1905. The idea seems to be that the gradual building up of superiority in the firing line will ensure success. Thus:

> The climax of the infantry attack is the assault, which is made possible by superiority of fire. The fact that superiority of fire has been obtained will usually be first observed from the firing line . . . The impulse for the assault must therefore often come from the firing line . . . Should it be necessary to give the impulse for the assault from the rear, all available reinforcements will be thrown into the fight, and as they reach the firing line, will carry it with them and rush the position.

---

[7] This and subsequent quotations are from Field Service Regulations Pt. 1 1909, ch. VII.

The importance of concealed entrenchments and cross fire to cover open ground is stressed, but there is no sign that the authors realised the overwhelming strength given to the defence by the machine-gun. Machine-guns in fact get more mention under attack and counter-attack than in defence, and their limiting factor is taken to be that ' rapid fire cannot be long sustained owing to the expenditure of ammunition involved.' The artillery is seen as the decisive supporting weapon: ' localities which present difficulties to the infantry alone may, if converging and enfilade fire of artillery is brought to bear on them, be carried with comparatively little loss.'

The position of the commander in battle is one of the problems which was to prove most difficult in the Great War. The teaching in Field Service Regulations was borne out by events, but the following extracts show how different was the scope of operations contemplated from what was to come.

> During an engagement the position of a commander will depend a great deal on the size of the force he commands. With a small force it may be possible to exercise personal supervision, but with very large forces the commander-in-chief should usually be well in rear, beyond the reach of distraction by local events, and in signal communication with his chief subordinates. Subordinate commanders should take up positions where they can obtain a good view of the area in which their commands are operating, and which admit of easy communication with their immediate superior and the units under their command.

In the advance the teaching was, ' When an encounter with the enemy is anticipated, it is advisable that the commanders of columns should be well forward, usually with their advanced guard.'

The writers of both editions obviously failed to foresee that war would one day be a contest between formidable forces in fortress conditions stretching from Switzerland to the sea, nor did they foresee the devastation and chaos which artillery fire to breach such positions would cause for both sides. Short of this there was no failure to read the lessons of modern fire power. All the lessons, still valid in 1969, that Henderson had learnt from the American Civil War, are to be found in the teaching. Mystify and mislead the enemy, attack him where and when he is at a disadvantage, but when you do attack, the utmost determination is essential to success. Possibly the stress on the methodical stages by which an attack should be developed and a failure to mention specifically that reinforcements should be directed to exploit success rather than to renew an attack that had failed, might be said to lead to false lessons. Against this however can be set the teaching that the strongest effort will always be made against the weakest part of the position. It is impossible therefore, to blame the pre-war teaching or tactics for the enormous casualties that were to be suffered in the succeeding years. The extent to which these casualties were the result of mistakes which could have been avoided is discussed later.

Field Service Regulations also dealt with training in peace. Gone were the days when, as under the Duke of Cambridge, the training of the soldier consisted only in drill and routine with occasional reviews and field days, while the training of the regimental officer was almost completely neglected. Now training was seen as the preparation of the individual officer and man for the duties he would have to undertake in war. The training of the officer was divided into ' practical training, that is field training and manoeuvres, and theoretical training by means of staff rides, essays, war games, lectures etc.'[8] The training of the non-commissioned officer was approached in the same way : ' while hardly less important than the training of the officer and non-commissioned officer is the adequate physical and intellectual training of the private soldier, who must not only be taught to march, ride, use his eyes, judge distance, shoot, swim, scout, etc but must also be able to comprehend the meaning and object of every movement he is instructed to carry out.' The military year was to be systematically divided into progressive stages; individual training, company, regimental, brigade, divisional and collective manoeuvres.

So much for the theory and what should happen. What did happen depended naturally on the zeal and skill of commanders

[8] This and subsequent references are from F.S.R. 1909 Part 2, ch. VII, sect. 158.

at all levels. An army organized to fight in continental war had been built up in less than ten years, but this is a short time for the new outlook to permeate every rank. There were a number of enlightened officers at the top and a good leavening of officers who thought out the problems of tactics and training. But there were plenty of commanding officers who could not look beyond their own regimental barrack square and for whom the drill and discipline they had learnt as subalterns was all that there was to know about soldiering. Field training did take a more important place than it had before the Boer War but it went little further than teaching the short rushes in extended order and elementary fire and movement which would enable the firing line to be built up preparatory to the assault. The day's field training ended with the assault, which was always held to be successful. Too much attention was given to the thought that if an assault was pressed with determination it was bound to succeed and too little to what was to be done if it did not, or if it succeeded in one part but not in another The problem of consolidation and exploitation were dealt with in the text books but not practised.

Field Service Regulations did not mention the cavalry charge but, notwithstanding the lessons of the Boer War the sword was still the primary weapon, and regiments still trained on the understanding that to deliver a mounted charge was their ultimate purpose. Nevertheless due to fine horsemanship and horsemastership, learnt as much in the hunting field as at regimental training, British cavalry were admirably fitted for their role of reconnaissance, scouting, and protection, which they were to carry out so well in the mobile operations of 1914. They were to prove less fitted for a place in the break-through after the long years of waiting.

There were thus deficiencies in training and outlook which the teaching of Henderson and others and the experiences of the Boer War had not completely repaired. Yet the Boer War had had a most salutary effect on British field work; both their minor tactics and their musketry were in 1914 superior to the French and the Germans who both still inclined to advance in mass. In equipment the British did not fare so well. There was probably little difference in field artillery, although the French 75mm gun was the best, but the Germans had far more heavy artillery and an excellent 5·9 howitzer. The Germans too were excellently and lavishly equipped with some of the refinements such as a light field cable, hand grenades, and binoculars. The greatest German superiority of all was in machine-guns. Not only did they have more but they had factory-filled belts (the British had to be filled by hand) and their rimless ammunition gave fewer stoppages.

In the training of senior commanders and staff officers the British had made great strides, although nothing could make up for the lack of actual practice in handling large bodies of troops, such as was a commonplace in the much bigger German and French armies. Some insight into the theoretical training actually carried out is found in a study of the last Aldershot Command Exercise[9] held before the Great War. The objects of this exercise, directed by Haig, the G.O.C.-in-C., were: first the ' development of quick mutual understanding between Commanders and Staffs in First Army so that in the event of war co-operation on sound lines, and in accordance with Field Service Regulations principles, may be assured between each part of the Command in the Field '; and secondly, ' The training of the staffs of the higher units in the duties that will devolve on them in war.'

The participants in the exercise were divided into syndicates each under a Brigade Commander. Before the exercise preliminary work was required. This consisted in making an appreciation of the situation as by the Commander-in-Chief of the Expeditionary Force and in drawing up instructions for the Covering Force. Maxse, the Commander of 1st Guards Brigade, was a syndicate leader. Haig's comment on his work is a valuable pointer to Haig's views on the relationship of a commander in the field and his subordinate commanders. He wrote: ' While on the one hand it is right not to interfere with the initiative of subordinate commanders, on the other hand *you as C.-in-C. the Expeditionary Force must take your share of the responsibility.*' (Haig's italics.)

The great military writers all tell us that the principles of war are unchanging. They are probably right, but those who train armies have more to do than to expound the principles.

[9] Full papers of this exercise are included in the Maxse Papers in the Imperial War Museum.

They must teach their application in the conditions of the day, but they must also foster the enquiring mind which will adapt them to the changes in conditions and weapons which the very impetus of war itself will accelerate. Whatever was to come it can be said, without fear of contradiction, that the British Expeditionary Force was the best trained and best equipped army that had ever left these shores. It could bear comparison too, with any army, allied or enemy, that took the field in 1914.

Chapter 3

1914: THE ENCOUNTER BATTLES

The mobilization of the British Expeditionary Force was ordered on 4 August 1914, the day war was declared. By 20 August the concentration of four divisions, the cavalry division, and one additional cavalry and one additional infantry brigade, was completed on the left of the French Army, between Maubeuge and Le Cateau. The two other divisions were delayed in Britain to guard against an early attempt at invasion, but 4th Division arrived in France on 22 August and 6th Division on 9 September. The British concentration went on unhindered while the Germans were engaged by the Belgian Army and the fortresses along the Meuse. Liège held out from 5 to 16 August but Namur, where the Germans quickly brought up heavy howitzers, held out only from 20 to 22 August. The Belgian Army retired into Antwerp while the Germans, in accordance with their plan, swept on to envelop the French Army. Although air reconnaissance was used and the Germans had a slight superiority in the air, they had no clear idea of the British movements or intentions. There was a strong presumption that they would detrain at Lille or Tournai and co-operate with the Belgian Army.

The British had indeed considered disembarkation at Ostend and operations based on Antwerp, but the size of the force precluded independent operations far from the main French forces and there was no time to do anything but fit in with the arrangements made by the French. The instructions given to Sir John French, who on mobilization had been appointed Commander-in-Chief, stated that his command was entirely independent and that he would in no case come in any sense under the orders of any Allied general.[1] Nevertheless his task was to support and co-operate with the French Army and the size of his force gave no scope for independent action. On 20

---

[1] *Edmonds Military Operations France and Belgium*, 1914 (Macmillan), Appx 8.

August Joffre ordered the general advance in accordance with the French pre-war Plan XVII. The French had underestimated the strength and direction of the German advance through Belgium and their Fifth Army was forestalled on the River Sambre. The British Expeditionary Force, covering the left of this army, first came into action at Mons.

The fighting at Mons, the subsequent retreat and the fighting at Le Cateau were all carried out in accordance with the teaching discussed in the last chapter. The cavalry provided good information of the German strength and movements. The infantry in hastily occupied positions, but entrenched and well hidden, had checked superior numbers by the excellence of their rifle fire. So well grounded was the British soldier in his musketry practice of fifteen aimed shots in a minute, that the Germans greatly overestimated the number of machine-guns in a battalion—it was in fact only two. The artillery, trained to engage the enemy as they were disclosed and to rely on long range observation through binoculars, were much hampered by the nature of the country—industrial and coal mining. 'The officers had great difficulty in finding suitable positions for batteries or even for single guns, and were equally at a loss to discover good observation posts. The general policy followed was to push batteries or sections up to the infantry line for close defence.'[2] By this means the artillery played a most effective part in the defence. But it meant that they were often left in position under an annihilating infantry fire and only extricated, at the last possible moment, by the gallantry of their six-horse teams. The performance of the British Expeditionary Force in the trying conditions of a retreat was all the more praiseworthy in view of the fact that the large proportion of reservists—in the infantry it was sixty per cent—had had no opportunity to get hardened.

On the command side it is more difficult to estimate performance. French's decision to withdraw from Mons was inevitable and was in accordance wth Joffre's wishes. There is clear evidence that Kluck (German First Army) intended to destroy the British Army on the line Maubeuge-Valenciennes[3] and the withdrawal prevented this. The orders were clear and

[2] *Ibid.* p. 75.
[3] *Ibid.* p. 149.

1914: The Encounter Battles 47

issued in good time and the new position was already reconnoitred. But as so often happens the withdrawal did not go according to plan. Haig's I Corps on the right got away without difficulty but part of Smith Dorrien's II Corps was so closely engaged by the enemy as to be delayed, and then exposed by the withdrawal of I Corps on its right and the cavalry on its left. Smith Dorrien saw that it was too late to carry out the withdrawal already ordered and that he must first check the enemy advance. With the co-operation of Allenby's Cavalry Division and 4th Division, which had just arrived, he fought the Battle of Le Cateau. Smith Dorrien had been in touch with G.H.Q. by telephone and had got somewhat unwilling concurrence in his decision to fight. Notwithstanding this knowledge at G.H.Q. no steps were taken to alter orders to I Corps, which carried on with its withdrawal thus creating a dangerous gap between the two Corps.

Despite the dangers of the situation the action at Le Cateau was most effective in allowing the B.E.F. to get away intact. In the first place the enemy suffered heavy casualties and lost a whole day. In addition they failed to find either exposed flank of II Corps and lost touch with the B.E.F. for the remainder of its retreat. Kluck worked on the preconceived idea that the British were facing east and withdrawing towards Calais. On 27 August, the day after Le Cateau, Joffre visited French at his headquarters. He was already planning his counter-stroke but he said it would be necessary to withdraw further than he had intended, to a line from Reims to Amiens. He asked that the British should withdraw stage by stage so as not to uncover the French Fifth Army. By the next night there was still a gap of some eleven miles between I and II Corps only lightly covered by cavalry, while the B.E.F. was some six miles behind the left of Fifth Army.

A similar gap—some fourteen miles—existed between the German First and Second Armies. Kluck's ignorance of the British line of withdrawal was one of the factors which caused him to acquiesce in Bulow's request to close in on Second Army, thus directing his line of advance east of Paris to envelop the French left. It was this change of direction which made possible the counter-offensive for which Joffre was awaiting the opportunity. In his orders for a further withdrawal, towards

Paris, issued on 30 August before Kluck's inward wheel had begun, Joffre made no mention of his proposed counter-offensive. He asked that the British should cover the gap between the Fifth and the newly created Sixth Army. French's reaction was to send a note in his own hand, saying:

> The new plan of retreat having been explained to me, I consider it absolutely necessary to inform you that the British Army will not be in a state to take its place in the line for ten days. I am short of men and guns to replace losses which I have not been able to ascertain exactly owing to the uninterrupted retreat under the protection of fighting rearguards. You will understand in these circumstances that I cannot comply with your request to fill the gap between the Fifth and Sixth Armies, that is to say, on the line Soissons-Compiègne.[4]

French ordered the retreat to be continued on 1 September despite Joffre's plans and despite the repetition of Joffre's plea in the words, 'I earnestly request Field-Marshal French not to withdraw the British Army until we are compelled to give ground, and at least to leave rearguards, so as not to give the enemy the clear impression of a retreat and of a gap between the Fifth and Sixth Armies.'[5]

1 September was an important day in several respects, not least for several rearguard actions, including 1st Cavalry Brigade and 'L' Battery Royal Horse Artillery at Néry, in which the British showed themselves at least a match for the Germans. On this day, too, Lord Kitchener, disturbed by information that French seemed no longer willing to co-operate with the French Armies, arrived in France to see the British Commander-in-Chief. Kitchener made it clear that the B.E.F. must conform to the movement of the French Army, although it must at the same time act with caution in order to avoid being in any way unsupported on its flanks. By that evening the gap between I and II Corps had been closed, while the cavalry covered the gap between III Corps (which had been formed on 30 August to command 4th Division and 19th Infantry Brigade) and the

[4] *Ibid.* p. 245.
[5] *Ibid.* p. 249.

1914: The Encounter Battles 49

French Sixth Army. The B.E.F. was however still at least a day's march further south than the left flank of the French Fifth Army.

While the retreat continued British air reconnaissance discovered that Kluck was moving south-east to attack the Fifth Army and was thus exposing his own right flank to the B.E.F. and the French Sixth Army, now incorporated with the garrison into Galliéni's Army of Paris. By 6 September the moment for Joffre's counter-stroke had come. To look at the larger picture for a moment five of the German armies were now enclosed in a large sack between the fortresses of Verdun and Paris, with the sides of the sack held by troops ready to turn to the offensive.

Joffre made up his mind by the evening of 4 September and orders for the attack were sent out to all concerned that night. Both the B.E.F. and Fifth Army had already begun the next stage of the withdrawal before the orders were received so both were further south than Joffre intended when the advance began on the 6th. Both Galliéni and Joffre visited French on the 5th, the latter in order to ' beg in the name of France the intervention of the British Army in a battle into which he had decided to throw his last man.'[6] French left no doubt that he would give his wholehearted support and that his army would do all that men could do. Joffre's plan was that his two armies of the centre, Fourth and Ninth, should hold while Third Army on the right and Fifth, B.E.F. and the Army of Paris on the left would attack the two flanks of the enemy between Verdun and Paris.

In the event the Battle of the Marne involved no major actions; neither the British nor the French armies did more than push in enemy rearguards. The German retreat was an example of the psychological factor in war, the influence of the will of the commander. Kluck and Bulow as well as Moltke, at Supreme Headquarters, finding the position in which the right wing had placed itself, feared a trap and lost their heads. The crossing of the Marne by the B.E.F. on 9 September was important because it seemed to threaten the dangerous gap which still existed between Kluck and Bulow. Thus the Marne was decisive in ending the German advance and the threat to Paris, but it

---

[6] Muller, *Joffre et la Marne* (Paris, 1931), p. 105.

cannot be claimed that either the British or the French brought about the result by any significant fighting.

It was after the Marne that the real opportunity for decisive action occurred. In order to close the gap which had caused the German retreat Bulow was given command over Kluck's Army, but his fears for his own right flank together with Kluck's slowness to respond to his directions caused the gap to increase to thirty miles. The B.E.F. was in front of this gap, and the only hope of the Germans restoring the situation was to hold the line of the Aisne or the high ground north of the river. There was no indication in French's operation orders, which were issued daily between 9 and 14 September[7] that there was a gap in front of the advancing divisions. More serious than this no tactical guidance as to the ground to be gained was given, nor was there any indication of the necessity for speed to seize an opportunity which would otherwise pass. Joffre also was not as explicit as he might have been, although in his instruction of 10 September he did say: 'To confirm and exploit this success, the advance must be pursued energetically, leaving the enemy no respite, victory is now in the legs of the infantry.'[8]

There is no doubt that the capture of the Chemin des Dames ridge at this time would have been of overwhelming importance and might even have led to the end of the war in the west. The German armies were exhausted by the strain of their rapid advance through Belgium and France, combined as it was with a good deal of heavy fighting. Now there were signs that they were beginning to be demoralized by the retreat and the fears that their situation was worse than it really was. The Chemin des Dames ridge rises up from the Aisne and dominates the surrounding country rather like Salisbury Plain dominates Wiltshire, and on something like the same scale. The southern slopes are heavily wooded but observation from the ridge to the north is even better than towards the south. The River Aisne and the ridge offered the Germans almost the only chance to halt the allied advance and to get the eight days which Moltke considered necessary for rest and to bring up reinforcements and ammunition. It was along the ridge that the gap occurred and that Bulow was weakest; he had only three tired

---

[7] Edmonds, *op. cit.*, Apps 39-45 give the orders in full.
[8] *Ibid.* p. 364.

1914: The Encounter Battles 51

cavalry divisions to cover the thirty miles. His hope for the future lay in the newly constituted Seventh Army, made up of a corps set free by the surrender of Maubeuge on 7 September and two corps moved up from the south. The Seventh Army was put under Bulow who now had the three right flank armies under his command.

In the advance from the Marne the British divisions, screened by the cavalry, covered about ten miles a day and by nightfall on 12 September the whole B.E.F. was close up to the Aisne. No real attempt was made to anticipate the Germans at the bridges and only in one place was a resolute effort made to seize a crossing place by a surprise coup. This was in 4th Division where 11th Brigade, commanded by Hunter-Weston (of whom we are to hear more when we discuss Gallipoli), carried out a daring night attack to seize the Venizel bridge, which had been only partially destroyed. This division had marched thirty miles in pouring rain in the preceding twenty-four hours and its performance is an indication of what might have been achieved had a greater sense of urgency prevailed throughout the higher command. The remaining divisions secured crossings on the next day but it was only on the right of Haig's Corps that a footing was secured on the ridge overlooking the Aisne. The French had advanced no more quickly than the British, but it must be remembered that the full strength of Bulow's Army faced the French Fifth Army and that Manoury's Sixth Army faced Kluck and was not only concerned with keeping touch with the left of the B.E.F. but also with attempts to turn Kluck's outer flank. The B.E.F. was in touch with the French on both flanks but between 1st Division on the right and 4th Division on the left the B.E.F. was strung out and over-extended astride the Aisne. Possibly some opportunity still remained; if Haig, with XVIII French Corps on his right could get full possession of their part of the ridge, then a move in conjunction with Manoury's army north of Soissons could sweep the Germans from the ridge. But by mid-day 13 September the Seventh Army had begun to arrive in strength and by that afternoon the Allied opportunity had passed. Hard fighting on the 14th produced disappointing results, such little progress as was made was very costly. All five divisions of the B.E.F. were committed, and were too widely extended to allow a strong attack anywhere. The 6th Division, which did not arrive until 16 September, plus two more divisions and some medium and heavy artillery, would have been necessary for the weight of attack that was necessary at this stage. The Battle of the Aisne was now over and the Chemin des Dames feature was not to be occupied until almost three years later, in the Nivelles offensive of 1917, a battle so costly as to undermine the efficiency of the whole French Army.

It would not be just to criticize French and his senior commanders for failure to make the most of their opportunity without taking into account the situation as it seemed to them at the time. Their experiences at Mons and in the retreat had been a nasty shock and they had not yet fully regained confidence in their allies. Their men had been fully on the go for more than three weeks and, although first reinforcements had rejoined their units, most of the infantry battalions were still much understrength. The enemy had a much greater weight of artillery and the disparity was made worse by the fact that our losses in field-guns had not yet been made good. As was to be seen when the Germans had occupied the commanding ground, our army was not strong enough in divisions or in guns to make an organized attack on the front allotted to it. Nevertheless the opportunity of forestalling the enemy had arisen; this was exactly the kind of situation for which our army had been trained and they ought to have been asked to do as much in the advance as they had done in the retreat. The Marne had been crossed without meeting any major resistance and French and his Corps Commanders should have recognized the ground that would give them the tactical advantage, and the vital importance of the time factor.

There is no doubt that if our commanders had been faced with this situation in a paper exercise they would have realized that the first requirement was to seize crossings over the Aisne, and that immediately this was done it was essential to get the Chemin des Dames Ridge. Our five cavalry brigades were still in good order and had proved themselves a match for the enemy. Instead of continuing in a protective role they might have been used, as was the teaching of F.S.R. discussed above[9]

---

[9] See page 38.

1914: The Encounter Battles 53

to seize this important tactical position and hold it until the infantry arrived.

Looking at the whole phase from Mons to the Aisne it can be asserted that the pre-war training methods and tactical doctrine were proved to have been sound. There are grounds for thinking that the choice of French as Commander-in-Chief was not a good one. He was not the dashing leader that some had thought him in the South African War. The weaknesses that had then appeared as well as the brilliance were real. Moreover, he had not kept himself at that pitch of perfection which his position at the head of the army demanded. Haig and Smith Dorrien, the two Corps Commanders that had borne the brunt of the Retreat, had shown themselves well up to their task, although Haig might with more initiative have remedied some of French's lack of drive in the advance to the Aisne. The staff had coped not only with the difficulty of a quickly changing situation, but with the evacuation of the advanced base at Amiens and the transfer of the main base from Le Havre to St Nazaire in the middle of operations. The regimental officers and the men in the ranks had, as a whole, risen to every demand made upon them. The basic training in musketry and marching paid a full dividend, which was all the more remarkable when the high percentage of reservists is remembered. The discipline, the courage, and the ability to adapt themselves to circumstances was all that those who know the British Soldier would expect. Altogether the performance of the B.E.F. in all ranks from top to bottom could bear comparison with any of those armies that went to war in 1914.

Chapter 4

THE SEARCH FOR A WAY ROUND

During the Battle of the Aisne Joffre had been concerned as much with turning the right flank of the German Army as he had been with exploiting the gap between Kluck and Bulow. As the Germans closed the gap both sides turned their attention to the outer flank, and both the French and Germans drew on their reserves and on their armies south of Verdun to create new armies in what was called, somewhat erroneously, the ' Race to the Sea.' Such British Forces as were available were landed at the Channel ports for a move towards Antwerp, where the Belgian Army was still holding out. First the Marine Brigade and then two more brigades of the Royal Naval Division were landed at Dunkirk, then at Zeebrugge 7th Division and 3rd Cavalry Division, constituted from units drawn from South Africa, India, and the Colonial Garrisons. With the extension of the front towards the sea and the importance to the British of the Channel ports it was obvious that it would be advantageous to have all the British forces again concentrated on the left of the French. The move of the B.E.F. from the Aisne and their concentration in the Ypres area was completed by 19 October. But by this time Antwerp had fallen and there was a French contingent and the Belgian Army continuing the British left from Ypres to the sea.

Ypres was the last mobile battle before the two sides settled down to trench warfare. The Allies had intended to take the offensive towards Menin, but, as Galliéni said, the Allies were always twenty-four hours and an army corps behind the enemy. As the British forces arrived they were thrust into the line to hold the German offensive designed to capture Calais and Dunkirk. Lack of any major reserves prevented the British from wresting the initiative from the Germans. For the B.E.F. it was a question of hanging on against superior numbers and of patching up the parts of the front that were beginning to crumble. In this situation Haig, in particular, showed his

The Search for a Way Round

competence in the use of local reserves. He employed his cavalry effectively as mobile reserve and when possible he withdrew units from parts of the front where risks could be taken, for bold counter-attacks in vital sectors. The superior musketry of the British infantry, together with its discipline and its dogged courage, enabled the B.E.F. to rise to its task. Possibly they were a little slow to develop the trench system that would have made their task easier. The reason was partly the shortage of picks and shovels, many having been lost in the retreat and in night work and withdrawals in the early part of the Ypres battle. Even worse was the shortage of wire. But an important reason also was the attitude of mind towards the defensive. Defence was considered a temporary expedient; they should soon be advancing again and the construction of a position for prolonged defence was considered a waste of time.

Ypres was the graveyard of the old regular army. In the infantry battalions that had fought from Mons to Ypres there remained only about one officer and thirty men of those who had landed in August. The battle was also the graveyard of the hopes for a short war. Kitchener, who had become Secretary of State for War at the outset, was one of the few who had not pinned his hopes on a short war. On 5 August he had stated that Britain must be prepared to put armies of millions into the field and to maintain them for several years. But despite his far-sightedness and his powers of organization Kitchener was in most ways unsuited to his post as Secretary of State, and would have been equally so as C.I.G.S. He undertook to raise, train, and equip new armies and to mobilize the nation's industry for war as well as to direct the conduct of British strategy. But he was quite incapable of using or developing the machinery of government. Neither the Defence Committee nor the War Office was allowed to play its essential part in the prosecution of the war. He also neglected altogether the resources of the Territorial Force for raising his new armies and he did nothing to introduce a scheme by which experienced officers and non-commissioned officers of the regular army were brought home to train new units. Kitchener spoke for the War Office, the C.I.G.S. became a cypher, and the General Staff at the War Office ceased to function as such. To make matters worse the Committee of Imperial Defence was superseded in November by the War Council. This was a cabinet committee attended by those cabinet ministers concerned, some additional members, and service chiefs summoned haphazard. As it was a cabinet committee, the generals and admirals only gave opinions when called upon to do so. As Asquith usually called only on Kitchener and the equally dominating Winston Churchill, First Lord of the Admiralty, others seldom got a look in.

The fighting at Ypres indicated the degree to which stalemate had been reached on the western front. Churchill's fertile imagination, searching for a way round, first suggested an operation with naval support to recover Ostend and Zeebrugge, and to clear the coast up to the Dutch frontier. French was favourable to the idea but demanded some fifty Territorial divisions, heavy artillery, and a liberal amount of ammunition. Joffre, whose co-operation was essential, thoroughly disapproved of the idea. Kitchener at first approved in principle, but later strongly opposed what he described as the 'wild-cat scheme' discussed between Churchill and French.

The situation in Russia soon turned the mind of the Cabinet and the War Council to matters further afield. The Allies had placed great hopes on the advance of the 'Russian Steam-Roller' and indeed at first all had gone well for her. Her armies advanced deep into East Prussia and into Galicia. In East Prussia the appointment of Hindenburg, with Ludendorff as his Chief of Staff, enabled the Germans to turn the tables by the brilliant victory at Tannenberg at the end of August. The Russian successes continued longer against Austria but the Germans switched an army and, again by the genius of Ludendorff, the Russians were defeated and forced back to the Vistula. The Russian difficulties were increased by the entry of Turkey into the war against the Allies on 5 November. In January 1915 the Russians appealed to Britain for action against the Turks to relieve the pressure against them in the Caucasus. Out of this appeal and the growing conviction of many that the deadlock on the western front could not be broken arose the controversy between Easterners and Westerners. The hopes of the Easterners were twofold. They still placed great faith in the capacity of Russia, although in fact after Tannenberg the Russian Army did not again set foot on German soil. They also believed that Germany could best be

The Search for a Way Round

struck through her allies, a policy known as 'knocking away the props.' Whatever the strength of the case for the Easterners this phrase was misleading; Germany was the prop and stay of the Triple Alliance. Her allies were more often an encumbrance than a prop. Fundamentally the Easterners based their strategy on Bacon's dictum that he who commands the sea is at great liberty and can take as much or as little of the war as he will.

Out of this aspect raged a controversy which lasted until March 1918. The Easterners believed that substantial withdrawals could be made from the western front without danger. The Westerners believed that the war could be won only by the defeat of the German Army, and they believed that this could be achieved only on the western front. They also believed that if the British weakened their forces in the west for adventures elsewhere they themselves ran grave risk of defeat. At no stage did the Westerners give up immediate hopes of overcoming the tactical deadlock. Their views on the means by which victory was to be achieved altered from time to time (as will be shown in succeeding chapters) but the next major offensive was always to be the last, or at worst the last but one.

Not all the Easterners favoured the same theatre for decisive operations. Churchill was for knocking out the Turks by an assault on the Dardanelles and the capture of Constantinople. Kitchener inclined towards an attack against Turkey at Alexandretta. Lloyd George was in favour of a move through the Balkans into Austria. Operations against the Turks either at Alexandretta or the Dardanelles would have the advantage that we should be saved the defensive operations necessary to safeguard the Suez Canal and the Anglo-Persian oilfields. The prospects against Austria varied as time went on. At first it was a question of saving Serbia, but the entry of Bulgaria into the war on 13 October 1915 killed this hope and Serbia was overrun by the end of the year. This did not end the efforts to bring in Greece and Rumania, or Lloyd George's interest in operations based on Salonica. Italy who declared war against Austria in May 1915 also offered a field for a British share in operations against Austria.

There were both political and military reasons why the British found it difficult to decide on and pursue a coherent war policy in 1915. They became involved in a political area in South Eastern Europe where long seated rivalries and territorial claims and counter-claims bedevilled every issue. A friendly gesture towards one Balkan country was in itself a hostile move against another. In August 1914 Greece, in which Monarchy and Government were then united, had offered to put all her armed forces at the disposal of the Entente. This offer was refused because the British were then working to keep Turkey friendly, or at least neutral. Then when Turkey joined their enemies, they hoped to unite all the Balkans on their side so as to save Serbia and crush Austria. However, the claims of Greece and Bulgaria were incompatible. By this time two factions had developed in Greece, the King on the side of the Germans and Venizelos in favour of the Allies. Even that was not a straight issue, because Venizelos could have nothing to do with the Allies if they tried to bribe Bulgaria at the expense of Serbia and Greece, even though Serbia and Greece were later to be recompensed at the expense of Austria and Turkey.

Asquith, the Prime Minister, well understood the proper relationship between the government and its military advisers. In the House of Commons on 15 November 1915 he said:

> It is the duty of any government . . . to rely very largely on the advice of its naval and military counsellors. But in the long run a government which is worthy of the name, must bring all these things into some sort of proportion one to the other, and sometimes it is not only expedient but necessary to run risks and so encounter dangers which pure naval and military policy would not warn you against.

Nevertheless the business of the War Council was not well organized. With the headstrong Churchill and the omnipotent Kitchener as service ministers it was not always clear where the military advice came from. Moreover, both ministers and service chiefs often left the meeting without knowing what had been decided. A most important example of this was the meeting on 13 January 1915 at which the plan to force the Dardanelles by naval action alone was discussed. Asquith thought the Council had sanctioned the making of provisional

The Search for a Way Round 59

plans. Churchill went ahead on the assumption that the decision had been made and that only details remained to be settled.[1]

Such vagueness and want of precision were typical of the way British military policy was directed before the war and during the first year. The cabinet had decided on a continental policy in alliance with France without knowing what they were in for. The British foresaw a short war in which only their Expeditionary Force was involved on the Continent, and they made no plans beyond that. Kitchener's initiative gave them the possibility of large armies but there was no proper examination of the problem how best they could be used to win the war. France was interested almost exclusively in the western front, although, for political reasons not easily discernible, she took the initiative in the despatch of forces to Macedonia. In the early days France was happy for Britain to use a number of Territorial and new divisions in the Mediterranean but she wanted us to send all available experienced divisions to France and she was strongly opposed to the withdrawal of any forces already there. An examination of the problem would be the normal function of the General Staff at the War Office but for two reasons this staff had ceased to function. The officers from almost all the key appointments had been taken to make up G.H.Q. of the B.E.F. and, as has already been pointed out, Kitchener was the last man to delegate work properly or to make the staff machine function. Not only did he fail to use the General Staff to make the necessary reviews of the situation and to make plans, he did not even tell them the questions that were being discussed or the decisions that were made in the War Council. An illustration of the way Kitchener worked is the briefing of Ian Hamilton for command of the army in the Dardanelles operations. Until his arrival in Kitchener's room on 12 March Hamilton knew nothing of the project. Nor apparently did the C.I.G.S., Wolfe Murray, who, with the Director of Military Operations, had also been summoned. Kitchener's only instructions were: 'We are sending a military force to support the fleet now at the Dardanelles, and you are to have command.'[2] Hamilton was able to elicit from Kitchener the number of troops that would be made available to him,

[1] Aspinell-Oglander, *Military Operations Gallipoli*, vol. I (Heinemann), p.59.
[2] Hamilton, *Gallipoli Diary*, vol. I (Arnold, 1920), p. 2.

but nothing about the enemy and only a few out of date handbooks on the Turkish Army and an inaccurate map of the Peninsula. On the next day Hamilton was again summoned to receive his written instructions which Kitchener was still working on when he arrived. The two essential points that came out of the orders are that once the navy had entered on the undertaking it could not be abandoned and that the decisions whether and to what extent the army was to be involved were left to Hamilton. As will be seen, in these two decisions, or rather in Kitchener's failure to let his colleagues in the War Council and at the Admiralty know that this was the way he saw the campaign, lay the seeds of our failures to force the Straits.

To return to events as they unfolded themselves; at about the same time as the Russians appealed for help against the Turks there were two proposals to the War Council for a review of the use of the new forces which would become available. Both pointed to operations in the Mediterranean. Hankey, the Secretary, by direction of the Prime Minister circulated a paper on 28 December 1914 suggesting that the deadlock on the western front could be resolved only by some new method or device for capturing trenches protected by wire and machine-guns, or by striking at Germany's allies, preferably Turkey. Lloyd George's paper pointed out that some half million men would soon become available. They must not be squandered on the western front, which he considered impregnable. All that was required there were the French Army and a large British reserve kept near the coast in case of emergency. He suggested that the help of all the Balkan countries should be enlisted and that some 600,000 British troops based on Salonica or a Dalmatian port should be used for an attack on Austria. He also suggested a subsidiary landing on the Syrian coast to cut Turkish communications with Egypt. Whether or not one agrees with Lloyd George's plan no one could doubt his wisdom when he pointed out that preparations for such operations would take months and that 'expeditions which are decided upon and organized with insufficient care and preparation generally end disastrously.'[3] Neither plan was formally discussed by the War Council. Kitchener did, however, write to

[3] Aspinall-Oglander, *op. cit.*, vol. I, p. 50.

The Search for a Way Round 61

French and suggest that the feeling in London was that the deadlock on the western front could not be broken and that troops now becoming available, over and above those required to hold the present line, might be used elsewhere. French replied that it was not impossible to break the German line provided he were given sufficient guns and artillery.

At this time the B.E.F. consisted of eleven divisions and five cavalry divisions, of each of which two were Indian. In Britain there were two regular divisions, made up of units brought home from India and colonial garrisons, and a Canadian division. Two of these had already been promised to French. There were also eleven first-line Territorial divisions ready for service, although some of them were employed on home defence. The second-line Territorial divisions and the New Army divisions would not be ready for service for some months. There were in Egypt one Territorial and one Indian division; there were also an Australian division, a New Zealand brigade and three mounted brigades all of which would complete training in February.

On 28 January the War Council decided to force the Dardanelles by naval action alone. As preparations to this end went ahead it became apparent that, even if the naval action was successful, troops would be required to occupy the peninsula and perhaps Constantinople. On 9 February Kitchener told the War Council that if the navy required the assistance of land forces at a later stage it would be forthcoming. At the same meeting a report of Lloyd George's visit to Paris was received and it was agreed to offer one British and one French division to Greece on condition she would assist Serbia. On 15 February Greece refused so small an offer. The next day Kitchener agreed to mass troops on the island of Lemnos in case they were needed for Gallipoli. The 29th Division was ordered from England and the ANZAC Corps to stand by in Greece. The 29th, the last regular division, had now been under consideration for three different roles. It had been promised to French, earmarked for Salonica, and now put under orders for the Dardanelles. The embarkation orders were cancelled three days later on representations from French, supported by the French Government. Despite all this the War Office was not told that any large scale operations in Gallipoli were contemplated nor ordered to make any preliminary studies or plans. Kitchener was not the only person responsible for this omission; Wolfe Murray, the C.I.G.S., was present at all the War Council meetings. Without in any way excusing Wolfe Murray this is an example of the extraordinary hypnotic influence that Kitchener exerted—he, and he alone, would act.

On 19 February the bombardment began and the next day Churchill warned Kitchener that at least 50,000 men should be at hand, at three days' notice. On 9 March the ANZAC Corps sailed from Egypt, and on the 10th (the first day of the Battle of Neuve Chapelle) Kitchener decided to make available 29th Division. On 18 March the naval force in a second attempt to force the Straits suffered heavy losses. The men on the spot, Admiral de Robeck and Hamilton, decided between them that it would be best not to renew the naval attack but to carry out a combined operation. This decision was upheld by the War Council despite the protests of Churchill, who was not supported in his opinions by his admirals. Hamilton thereupon ordered his army back to Alexandria to prepare for an assault landing. All strategical surprise thus having been forfeited the first landings took place six weeks later at Cape Helles and the bay afterwards known as Anzac Cove.

The two most tragic elements in the Gallipoli story are the decisions, first to launch the naval attack without military support, and second to call off this naval attack in order to wait for military support. There seems little doubt that if Churchill and the admirals had known that Kitchener would make three divisions or so available they would have preferred to wait for a prepared combined operation. There is no doubt that the naval attack having gone as far as it did should not have been called off without a further effort. It is now known that the forts did not have sufficient ammunition to resist a renewed attack.

Despite all these mistakes in the inception of the campaign, the landings on 25 April came within an ace of success. If Hamilton had been able to force through the brilliant plan that was in his mind, the whole peninsula might quickly have been captured, but by May 1915 (just at the time the Aubers Ridge offensive was launched in the west) stalemate had set in. In August another brilliantly conceived effort to break the stale-

The Search for a Way Round

mate by a landing at Suvla Bay, combined with a new offensive from the Anzac position, was also allowed to run to waste. This failure ended all real hopes of gaining control of the straits and the capture of Constantinople. Some of the tactical lessons which come out of the failures are discussed in succeeding chapters.

Soon after the Suvla attack had petered out the French began to take up again the case for operations in support of Serbia. Their natural interest in the fate of an ally was reinforced by dissatisfaction with their subordinate position in Gallipoli, where they had two divisions, and by a desire to give an independent command to General Sarrail. Sarrail, who had just been superseded in command of the French Third Army, had a strong political following. It was from his followers that the Government and Joffre were under strong criticism for the failure of the offensive in Champagne and for the failure so far to do anything effective to help Serbia.

The certainty that Bulgaria was about to attack Serbia made Greece once more inclined to intervene. To help them the French intended to provide four divisions and asked Britain to provide two. By October Kitchener was beginning to realize the impasse at Gallipoli and suggested that the Suvla Bay position should be evacuated and two divisions made available to go to Macedonia. By this time Archibald Murray, former Chief of General Staff to French, who had become Deputy C.I.G.S. in January 1915 and C.I.G.S. in September, had begun to knock a little sense into the War Office. He had so far restored the General Staff to its proper function that Kitchener read out to the Dardanelles Committee (this was the existing War Council) an appreciation recommending that the role of troops sent to Macedonia should be restricted to protecting the Greek flank and communications into Serbia. Eventually it was decided to send one British and one French Division from Gallipoli, but not to evacuate Suvla Bay. On 5 October it was further decided to send three British divisions from France and that the French would provide two divisions and two cavalry divisions. But by the next day, as the leading elements of the first two divisions were landing at Salonica, Venizelos fell and Greece's attitude was once more in doubt. In the following days further doubts arose. It did not seem possible now to save Serbia. There were even fears that Germany would be able to move forces through in support of Turkey and that our army in Gallipoli would be in jeopardy. The French had none of these doubts and pressed for action to prevent the destruction of the Serbian Army. Joffre himself came over to 10 Downing Street to urge the importance of the projected operation. It is difficult to say to what extent this was an expression of Joffre's military opinion or whether it came from a realization that his own position as Commander-in-Chief in France depended on the acceptance of the operation. Robertson, then Chief of General Staff to French, writing after the war, said that practically all the French generals with whom he came in contact, including Joffre, Foch, and Pétain, showed in manner if not in actual words that they intensely disliked the project from the start.[4]

On 14 October the British Government decided to replace Hamilton by General Sir Charles Monro, who was instructed to report whether he recommended the evacuation of the Peninsula. Monro reported almost at once in favour of evacuation. But by this time Kitchener was much opposed to giving up the Peninsula and he refused to countenance it until he had seen the situation for himself. While in the Eastern Mediterranean Kitchener reverted to the possibility of a landing at Alexandretta. The General Staff at the War Office was now strong enough to quash a proposal for a further expedition while Gallipoli and Salonica were still on British hands. Kitchener visited Greece, where the political situation had further deteriorated. There was even a chance that Greece would be hostile. This danger was averted but the British Government, supported by the General Staff, was determined on the evacuation of Salonica. At a conference at Calais on 6 December, Asquith, Balfour (now First Sea Lord) and Kitchener met the French ministers and decided that if communications with the Serbian Army could not be made and maintained, the allied force would be withdrawn. French reaction to this decision was so strong that the War Committee, as the reconstituted War Council was now called, was prepared to go back on its decision. They authorized Grey and Kitchener to reach a settlement with the French Government. By 9 January 1916

[4] Robertson, *Soldiers and Statesmen*, vol. 2 (Cassell, 1926), p. 103.

The Search for a Way Round

the whole of the Gallipoli Peninsula was evacuated and the British were left involved in a commitment in Macedonia which had the support neither of the British Government nor of its military advisers because its original military aim had disappeared.

It was also an anxious time elsewhere in the Middle East. The necessity for protecting the Anglo-Persian oilfields had led in November 1914 to the despatch to Basra of a force controlled by the Government of India. A number of brilliantly executed operations had led General Nixon to undertake an advance far outside his defensive aim. The 6th (Indian) Division, under Townsend, continued its progress almost to the gates of Baghdad but there was no administrative capacity to allow it to fight so far from its base, or to subsist if it were checked for any length of time. The check came at the indecisive Battle of Ctesiphon in November 1915. The decision to fall back to Kut and there to stand siege was taken just at the time of the Calais Conference when the British were trying to extricate themselves from the expedition to Macedonia.

While the British star had been waning in the Eastern Mediterranean things had been going equally badly in France. Kitchener's new divisions had been committed in the Battle of Loos lasting from 23 September to 8 October. The attack had failed, as had the French offensive on its flank in Artois and that in Champagne. Between them they cost the Allies more than 300,000 casualties. For his mishandling of the Battle of Loos French was replaced by Haig on 25 November. At the same time Robertson replaced Archibald Murray as C.I.G.S.

It is not possible to judge the rival merits of Easterners and Westerners without answering the question of whether there was any real hope of breaking through on the western front and if so whether British resources were being properly used to that end. These are questions which will be discussed in succeeding chapters. Two things about the Easterners' case are however clear. 1915 was the year of their great opportunity. While Russia was still strong, the Serbian Army still in the field, and Turkey still isolated from her allies, there was an opportunity for Britain and France to use the command of the seas to bring available forces to bear on Germany's allies. In a properly managed campaign Constantinople could have been captured. Whether the British could have saved Serbia as well is doubtful but they could certainly have done so instead. Even if they had been able to do both the very arguments on which the Easterners based their case suggest that Germany had sufficient forces to maintain the deadlock on the western front and to take advantage of interior lines to support Austria. If the British had opened the Straits they would have had access to Russia, but what could they have done with it? They had not sufficient ammunition and equipment for their own forces and it is doubtful if they could have made good the Russian deficiencies or have done anything to make their command in the field more effective. As between Constantinople and the saving of Serbia, the second would undoubtedly have been the more useful. It is, however, doubtful if either would have materially altered the course of the struggle against Germany.

Up to the time Robertson became C.I.G.S. he had not worked closely with Kitchener, but he had been able to see the results of Kitchener's arbitrary methods. Although Archibald Murray had done what he could to restore the proper position of the General Staff at the War Office, Robertson was not prepared to accept the position in which as C.I.G.S. he remained a military subordinate to a Field-Marshal acting as Secretary of State. He therefore set out on paper his views of the proper relationship between C.I.G.S. and Secretary of State. The essentials of Robertson's case were that all military advice to the War Council [sic] should be given through the C.I.G.S. and that orders for military operations should be signed by the C.I.G.S. under the authority of the War Council; the Secretary of State for War would thus be in the same position as any other minister who was a member of the War Council. Kitchener pointed out that Robertson's suggestions were unconstitutional since the Secretary of State was responsible to Parliament for the army. They agreed, however, that the C.I.G.S. should be the sole channel for military advice to the War Council and that all orders for operations to execute War Council policy should be signed by the C.I.G.S. on the authority of the Secretary of State. A paper to this effect was accepted by the Prime Minister.[5]

From this time onwards the controversy between Easterners

[5] *Ibid.* vol. I, pp. 168-170.

and Westerners took on a new complexion. Robertson was a convinced Westerner, Kitchener's powers were waning, and the German offensive against Verdun from 21 February to 29 April 1916 turned men's eyes to the western front and the necessity for relieving pressure on the French. Then on 5 June Kitchener, on his way to Archangel for a visit to the Czar, was drowned. Lloyd George became Secretary of State for War. The changed situation had not altered Lloyd George's views. He soon pointed out to Robertson that our fifty divisions on the western front could not always hold their own against the German thirty and that no substantial victories had blessed the six months in which the General Staff had fulfilled its proper role.[6] Nevertheless the incipient struggle between soldier and statesman did not yet cause serious controversy. Lloyd George was busy with political activities which led to his becoming Prime Minister in December.

Just at the time that Lloyd George became Secretary of State for War the Battle of the Somme began in the west and in the east the Russians achieved a surprising and extensive success, destined to be their last in the war. Brusilov in an offensive against the Austrians overran the greater part of Eastern Galicia and captured 350,000 prisoners. The British had been trying to persuade Rumania to enter the war since the beginning of 1915. Not unnaturally the fate of Serbia had deterred her but this offensive gave her the opportunity to deal with Bulgaria while Austria was fully engaged elsewhere. An attack now by Rumania would also give just the opportunity the Allied Army in Macedonia needed without requiring the diversion of other troops. Bulgaria might well be knocked out of the war and Germany's route to the Black Sea closed. Rumania hesitated, her eyes were on Transylvania and she thought that she could limit the war to one with Austria-Hungary alone. It was not until 27 August that Rumania declared war and by that time the Germans had come to Austria's assistance and Brusilov's offensive had been pressed beyond hope of further success. In this and abortive offensives further north the Russians lost a million men. In a brilliant offensive under the German General Mackensen, the Central Powers turned on Rumania and by mid-December almost the

[6] *Ibid.* vol. I, p. 176.

whole country was overrun. Lloyd George has already been credited with the conception of a Balkan operation which might well have succeeded in January 1915—but there is no evidence that while he was Secretary of State he did anything to save Rumania, nor indeed in the light of her actions was there anything that could have been done in time. He did, later, make much of what the British ought to have done. In a paper prepared for use by Asquith at an Allied Conference on 15 December the whole story of the opportunities missed by the British in Macedonia in 1915 was set out and attributed also to 1916. Among other things, Lloyd George said:

> The Salonica Expedition launched in time would have saved Serbia and given us the Balkans. At best all that can be said for it now is that it is holding 250,000 Bulgarians and Turks with a force which is nominally at any rate double that number . . . General Milne's figures show that the aggregate number of Allied rifles available does not much exceed 100,000. The equipment in guns and transport of these troops is ludicrously inadequate even for the modest role which it is supposed to play. Neither General Foch nor Sir Douglas Haig would ever dream of attacking the tiniest Somme village defended by a single German regiment with the guns and ammunition General Sarrail and General Milne have at their disposal for the storming of over 200 miles of the strongest positions in Europe held by over 200,000 of the finest infantry.[7]

Lloyd George did not draw the logical conclusion from his own words that the missed opportunities in Macedonia could never return.

During 1916 there were other important developments too on the outer ring of the war. The War Office had taken over control of the operations in Mesopotamia but the attempts to relieve Kut had failed. On 25 April the garrison surrendered. The British Government did not, however, give up hope of capturing Baghdad, and General Maude the new commander was instructed to improve communications with a view to a

---

[7] *War Memoirs of David Lloyd George* (Nicholson & Watson, 1933), pp. 920-21.

The Search for a Way Round 69

renewed advance. Maude, one of those men of boundless energy who liked to do everything himself, set about his task with great ability. By December Maude had so far improved the health, morale, and training of his force that he was able to turn to the offensive. The capture of Baghdad on 11 March 1917 was the first major Allied victory of the war.

When he was replaced as C.I.G.S. Archibald Murray had been given command of the Egyptian Expeditionary Force with the primary role of defending the Suez Canal. He had quickly seen that the passive defence of the Canal was wasteful. He saw too the administrative preparations that were necessary for a forward policy and started on projects for a railway and water pipe line across the Sinai Desert. By the end of 1916 the whole of the Sinai Peninsula was in British hands and a well administered force lay on the borders of Palestine. The Turks had been defeated in several battles and Murray was beginning to think of the capture of Gaza. In 1916, too, the Arabs in the Hejaz revolted against the Turks, overthrew them in Mecca and laid siege to Medina. A young captain, T. E. Lawrence, succeeded in reaching the camp of Prince Faisal outside Medina, and in persuading him that the proper strategy for the Arabs lay not in direct attacks on the Turks but in keeping them dispersed and isolated and thus retaining the opportunity to attack their communications.

Just before Lloyd George became Prime Minister, Joffre, in consultation with Haig and Robertson, had decided on the Allied strategy for 1917. This was nothing but a repetition of the Somme battle; the British and French would attack astride the Somme, followed fifteen days later by a French attack north of Reims. The British were to bear the main share of the attack. Lloyd George's anger rose at the thought of the dreadful casualties on the Somme. He did not believe for one moment the soldiers who told him the battle had come within an ace of breaking the spirit of the German Army. After the war Hindenburg and Ludendorff confirmed that it had.[8] When he became Prime Minister Lloyd George lost no time in renewing proposals for an attack on Austria. This time it was on the Italian front, where little had happened in

---

[8] Ludendorff, *My War Memories* (Trans., 1920), vol. I, pp. 276-8, 307.

1915 and 1916. The frontier had both tactical and strategic disadvantages for an Italian offensive. The mountainous country was easily defended, while any offensive from Venetia (which offered the most favourable line of advance) was liable to be cut from the Austrian Tyrol and the Trentino. Nevertheless the Austrians had plenty to occupy themselves elsewhere and generally the Italian failure had been attributed to lack of heavy artillery. At a conference in Rome in early January Lloyd George suggested an attack by combined British, French, and Italian forces through the Julian Alps to Laibach (the Ljubljana gap of the Second War) and Vienna. Neither the British General Staff nor any of the Allied staffs had heard of this plan before. The account of the conference given by Lloyd George[9] shows only too clearly how misunderstandings between soldiers and statesmen and between allies arise. Lloyd George suggested that French and British guns should be lent to the Italian Army. As they would be required for the offensive on the western front later in the year they could only be lent for a few months. In the ensuing discussion one can see that Cadorna, the Italian Commander-in-Chief, thought he was being beguiled into an offensive by the loan of artillery which would be taken away from him before his offensive began. Lloyd George just as firmly believed that Cadorna was refusing valuable assistance because he was unwilling to take the offensive. Briand was against the Italian project because the spring offensive in the west had already been agreed upon, so no definite decision was taken at Rome. The General Staffs of the various countries were, however, instructed to go into the possibilities both of assisting Italy if she were attacked and of an Allied offensive. One result of the subsequent staff talks was that a plan was prepared for the movement of six British and six French divisions to the Italian front should it later become necessary. These preparations were to prove invaluable in October when the Italians were heavily defeated at Caporetto.

For a time in 1917 Lloyd George was converted to a belief in success on the western front. One of the few advantages reaped from the Somme offensive was that it had given the French time to recover. In the autumn of 1916 Mangin's Corps, under

---

[9] Lloyd George, *op. cit.*, pp. 1413-51, 2275.

The Search for a Way Round

the direction of Nivelle, commander of the Army of Verdun, had regained most of the ground lost in the great German offensive. As a result Nivelle had succeeded Joffre as Commander-in-Chief in the west, although Joffre remained for a time as Commander-in-Chief of the French armies on all fronts. Nivelle was a dashing commander with a brilliant record. It was not only his military success that endeared Nivelle to politicians but his persuasive personality. He could speak English fluently (he had an English mother) and he captivated Lloyd George. Among the changes which Nivelle made in Joffre's plan was an extension of its scope and the relegation of the British from the primary to the secondary role. In this Lloyd George saw some respite from the enormous losses we had suffered on the Somme. Robertson's warning that a great increase in the French losses would in no way diminish those of the British went unheeded. Nivelle talked of a violent and sudden blow. The operation was to take only twenty-four to forty-eight hours and the success which was expected would give the opportunity to pass through a torrent of reserves to roll up the enemy line and destroy his communications. On the other hand if the attack were not successful it could be broken off. Here again Robertson foresaw that once the battle had been engaged it could not be broken off until every effort had been made. Lloyd George listened neither to Robertson nor to Haig, who was even more doubtful about Nivelle's concept of a single artillery bombardment to cover the whole depth of the enemy position, that is nine to twelve trench systems covering from five to twelve miles. This time it was Lloyd George who was convinced of the possibility of a breakthrough. Robertson was told to send a special instruction to tell Haig that the Government attached the greatest importance to the plan agreed with Nivelle being carried out in letter and in spirit. Then in collusion with Briand and Nivelle, and without telling Robertson, Lloyd George undertook to make Haig subordinate to Nivelle in every way as if he were a French Army Commander. This almost led to the resignation of both Haig and Robertson and certainly removed any vestige of hope that either of them would ever trust Lloyd George again. Both soldiers sunk their personal feelings and with the assistance of Hankey, secretary to the Cabinet, a working agreement was drawn up giving Nivelle strategic direction of the B.E.F. for this operation only.[10]

Nivelle's plans were ruined by the German withdrawal to the Hindenburg line, about which he had been warned both by Haig and by his own subordinate commanders. Nevertheless he refused to adapt his plans to the new situation. The British attack at Vimy Ridge and Arras went in on 9 April. On the first day an opportunity for a real break through was created, but, as will be recounted, the opportunity was let slip. A secondary French attack at St Quentin went in on 13 April and the main attack on the Aisne, including the Chemin des Dames Ridge, on 15 April. Both attacks failed and the French Army was involved in casualties which for the time being left it disillusioned and broken in spirit. From now on, even to a greater extent than before, the British Army had to take the main burden on the western front. Haig prepared to take up his burden by a summer offensive to clear the Belgian coast. Fears of a new German offensive towards Calais and a desire to end once and for all the threat to their shipping in the Channel had for months drawn British attention to the Ypres sector to an extent which almost amounted to an obsession. Haig's desire for this operation was so strongly supported by the Admiralty and by some members of the War Cabinet that Lloyd George allowed preparations to go ahead; he could do no more than withhold final approval. He then reverted to his idea of an Allied offensive in Italy. Robertson strongly resisted this alternative and by the third week in July he was able to report to Haig that his operations had the whole-hearted support of the War Cabinet. Lloyd George had, however, been able to get his colleagues' support for an offensive in Palestine, where they decided to replace Murray by Allenby. In an attack on Gaza on 26 March 1917, Murray had been on the brink of succeeding, but owing to an unfortunate series of errors the battle had been called off just in the moment of victory. A second attack was made three weeks later, but by then all surprise had been forfeited. His work as C.I.G.S. and in Palestine showed Murray to be a most capable organizer and a brilliant thinker, but as a general he had one unforgivable

---

[10] A detailed and largely first hand account of this incident appears in Spear's, *Prelude to Victory* (Cape, 1939), ch. 8-10.

fault, he was unlucky. He was rightly blamed for his failure to press his attacks to the uttermost, a quality in Haig that did not earn him the unstinted support of his political chiefs. Murray's foresight and administrative work in Palestine laid the foundations of Allenby's later successes, as Allenby generously admitted. Allenby was given additional troops by the withdrawal of two divisions and some artillery from Salonica. Robertson, who was continually pressing for the reduction of the Salonica commitment, agreed somewhat grudgingly to this change. He said: 'As divisions were apparently not going to be got away except on Lloyd George's terms, the General Staff could only acquiesce in their going to Palestine where they would at any rate enjoy a better climate and be under British control.'[11]

The Russian Revolution in March had led to the gradual withering away of any hope of Russian success and so removed the corner-stone of the Easterners' case. By the end of 1917 Russia concluded an armistice with Germany. On the other side of the balance sheet the United States joined the Allies in April 1917 and before the end of the year their expeditionary force was beginning to arrive in France.

Before Haig's main offensive began there was a striking limited success by the British Second Army at Messines Ridge. This was a classic example of siege warfare and of what could be achieved by careful preparations and abundant explosives and ammunition when a limited attack was delivered with surprise and ingenuity. The main attack was launched by Fifth Army on 31 July and was pressed through all the mud and rain of the wettest August in living memory. The Third Battle of Ypres, more commonly known as Passchendaele where the last phases were fought, petered out by early November without appreciable gains to show for the appalling casualties. The failure was partly due to the weather but the mud and chaos were little worse than might have been expected from the experience of the Somme. The Germans used much improved defensive tactics and it is probable that their army was less near breaking point than it had been in 1916. The only real gain was that Pétain had been able to take full advantage of the respite to restore the French Army. As has been shown

[11] Robertson, *op. cit.*, vol. II, p. 143.

Lloyd George disliked the offensive from the first, but the War Cabinet had approved it and the question is how far a politician is justified in interfering in a military operation. A government must either support a commander or must replace him by someone they can trust.

In later November on dry downland further south, the Battle of Cambrai showed what could be done with a surprise tank attack. Over 300 tanks took part and all but the last German line were broken. In a counter-attack on 30 November the Germans regained almost all they had lost. The lessons of this most important battle, and of Third Ypres, will be discussed later.

While the Passchendaele Battles were in progress the Italians had suffered a grave defeat at Caporetto. This brought out the weakness in the Italian geographical position and necessitated a withdrawal to the River Piave. In accordance with the plans which Lloyd George had insisted should be drawn up, one British and one French army corps were despatched from the western front.

During October and November, while British and French assistance to Italy was being discussed, the three allies decided to set up a Supreme Allied War Council at Versailles ' charged with the duty of continuously surveying the field of operations as a whole, and by the light of information derived from all fronts and from all governments and staffs, of co-ordinating the plans prepared by the different General Staffs, and, if necessary, of making proposals of their own for the better conduct of the war.'[12] Lloyd George was insistent that the ' Permanent Military Representative' should not be national Chiefs of Staff. Accordingly Wilson was appointed for Britain. Foch was the French representative and was succeeded as Chief of Staff by Weygand, his trusted disciple and chief staff officer. Cardorna, the superseded Commander-in-Chief was selected as Italian representative.

The end of 1917 would have been black indeed if Lloyd George had not got his ' Christmas present for the British people', the capture of Jerusalem. In October Allenby had achieved a brilliant victory at Beersheba and Gaza. All the administrative difficulties were overcome and the pursuit boldly

[12] Edmonds, *op. cit.*, 1918, vol. I, p. 31.

The Search for a Way Round 75

handled. The Turks fought hard and foiled Allenby's first attempt on Jerusalem but on 9 December, as a result of a second attack in difficult conditions of continuous driving rain, the city surrendered. This success, following on the earlier success in Mesopotamia, confirmed Lloyd George in his determination that the final destruction of the Turks should be the principal aim for the spring of 1918. There was another difference of opinion between Lloyd George and Robertson. Robertson argued that the armistice with Russia would give the Germans a preponderance in the west which they might well exploit before the arrival of the American forces could alter the balance. Allenby's next campaign might give us Damascus or even Aleppo, but these were hundreds of miles from the heart of Turkey. Turkey would not give in while she saw Germany successful in the west. Lloyd George retorted that the new German superiority of numbers in the west was less than that which the Allies had had for two years, and the Allies had been unable to make any real impression on the German line. It was mainly on this issue that the final break between Lloyd George and Robertson came. On 19 February 1918 Henry Wilson became C.I.G.S. All that had happened since the end of 1914 seemed to suggest that Robertson was wrong but the events in the west were to show how much sense and knowledge there was in his arguments during this last difference of opinion. During 1918 there was none of the stalemate of the earlier years but a series of serious breaks through the Allied line which brought them to the brink of defeat, then the turn of the tide—as in 1914 at the Marne—followed by a series of great British offensives which caused Germany to sue for peace.

Henry Wilson was the exact opposite of Robertson. He was in many ways a brilliant soldier and he had the art of getting on with politicians, which Robertson so clearly lacked. He had a nimble wit and a ready tongue which he often allowed to run away with him. For this reason he was not altogether trusted in the British Army and he had a reputation for intrigue. He was a fluent French speaker and he had as high a reputation with French soldiers as with politicians. Lloyd George hoped that with Wilson as C.I.G.S. he would get support for operations on such of the eastern theatres as still existed, but events took charge so that political and military energies were directed principally to the western front.

One of the results of the German breakthrough on 21 March was to bring the Allies to a unified command. The German attack on a forty-three mile front Arras-St Quentin-La Fère almost succeeded in reaching Amiens and driving a wedge between the French, with their fears for Paris, and the British, with their preoccupation with the channel ports. A second blow in the Lys sector achieved an unexpected degree of success and almost reached Hazebrouck. Largely at the instigation of Haig, Foch was appointed Generalissimo on 14 April. This was not the kind of arrangement so strongly opposed by Haig and Robertson which subordinated the British Army to the French —as had been attempted by Nivelle. The French Army under Pétain and the British under Haig both worked under the strategic direction of Foch. Foch himself took a leading part in the French counter-stroke on the Marne in July, but it had as its basis a cleverly fought defensive battle by Pétain. After this it was Haig who took the lead. There were many both French and British who believed that the best strategy was to wait until the arrival of the American armies would allow the knock out blow in 1919.[13] Haig believed that with a determined offensive we could end the war in 1918. The first attack was launched at Amiens on 8 August, described by Ludendorff as 'the black day of the German Army.' The Australian Corps in particular showed greatly improved tactics and technique, and the improvement was shown elsewhere as new attacks were driven home, first to turn the German position beyond the Somme, then further north at Arras, and finally in the old battlefield of Cambrai. By 3 October Ludendorff had persuaded the Kaiser that the war was lost and negotiations for an armistice had begun.

There were successes further afield too. After all the wasted years spent in frustration in unhealthy surroundings with no real military aim, the Allied army in Macedonia sprang to life and Bulgaria was the first enemy knocked out of the war. In strategic conception and tactical execution this last phase of a wasted campaign is as brilliant as any of the war. Franchet d'Espèray, who had become Allied Commander-in-Chief in

[13] Twenty-five divisions had arrived by July 1918.

The Search for a Way Round

June decided to do what the enemy considered impossible. He planned to use his six Serbian divisions, eager, hard and well trained in mountain warfare, supported by two French colonial divisions, to break through where the outer rim of the enemy defences was in the most difficult country and therefore held in the least depth. The attack, supported by secondary attacks by British and Greeks on the Dojran front to the east, and by the French to the west, began on 14 September. In three days the main attack had broken through to the River Vardar thus reaching the junction of rail and road communications to Skopje and into Central Europe. There the Bulgarian resistance completely collapsed and an armistice was signed on 29 September.

In Palestine, despite the reductions which he had suffered to reinforce the western front, Allenby achieved decisive success at Megiddo in September. In concert with the Arabs the pursuit was pushed on to Damascus and Aleppo. On 30 October the Turks capitulated.

The Italians, with British and French assistance, retrieved their earlier disaster by overwhelming success at Vittorio Veneto with the result that the Austrians concluded an armistice on 30 October. Thus all Germany's allies ended hostilities before Germany herself concluded the armistice of 11 November. But by no stretch of imagination can it be said that the cause of Germany's downfall was the collapse of her allies. Rather was it that because Germany showed herself on the point of defeat there could no longer be the hope of the usual move to save them and thus no purpose in their continued resistance.

Although one can dismiss the validity of any suggestion that the war could have been won without the defeat of the German armies in the field and one must agree that it was only in the west that these armies could be faced effectively, one cannot unreservedly approve the line that the military advisers to the government took in the many controversies. In particular the British missed opportunities against Austria and Turkey in 1915. The final success in Macedonia showed what might have been gained by keeping the Serbian Army intact in the field, a task well within the Allied capacity in that year. Asquith's lack of the power of decision and his failure to bring to a head the many excellent and far reaching plans brought before him were partly to blame. He can be excused to some extent because so much was expected of and left to Kitchener. Asquith's understanding of the theory of strategy and the extent to which he trusted Kitchener are shown by this entry in his diary:

> There are two fatal things in war. One to push blindly against a stone wall, the other to scatter and divide forces in a number of separate and disconnected operations. We are in great danger of committing both blunders. Happily K is a good judge in these matters—never impulsive, sometimes inclined to be over-cautious, but with a wide general outlook which is of the highest value.[14]

Kitchener's failure to reorganize an effective General Staff when it had been whittled away to form G.H.Q. of the B.E.F. was only part of the trouble. Kitchener himself was a curious blend of vision and blindness. Lloyd George perhaps best sums it up when he says of him: 'He was like one of those revolving lighthouses which radiates momentary gleams of revealing light far out into the surrounding gloom and then suddenly relapse into complete darkness. There were no intermediate stages.'[15] Kitchener and Robertson together were, surprisingly, a good combination. Undesirable as it is to have a soldier as Secretary of State Lloyd George, Kitchener, and Robertson together in 1915 might have achieved much. But by the time Robertson became C.I.G.S. Kitchener had lost most of his influence and some of his grasp of the problem. By that time, too, it was too late to do anything really effective outside the western front. Every expedition or reinforcement had to be judged coldly, as Robertson had done, as a diversion from the vital effort. Neither Lloyd George nor Robertson can be forgiven for their relationship with each other. They ought either to have parted or to have worked properly together. Lloyd George was too impatient and too intolerant of the blunt soldier of few words to take his views on their merits. Robertson was not sympathetic enough to politicians to state his views on their ideas fully and patiently, although he did so admirably

[14] Asquith, *Memories and Reflections*, vol. II (Cassell, 1928), p. 57.
[15] Lloyd George, *op. cit.*, p. 751.

The Search for a Way Round 79

but bluntly on paper. Another and possibly the worst cause of the differences was Robertson's relationship with Haig. Haig's responsibilities were solely in the west and he rightly devoted all his energies to gaining the victory there. He must be judged later in the light of the tactical problem there. But Robertson was in a superior position to Haig. It was certainly his business to shelter Haig from the vicissitudes of policy making and to see that he was given what was essential to fulfil his task. It can, however, reasonably be argued that Robertson went further than this and that he acted rather as Haig's subordinate representing his views and only his views to the Prime Minister. Only once or twice, all too rarely, is Robertson seen taking an independent line with Haig. It is tempting to look ahead and to compare the relationship between Robertson and Haig with that between Brooke and Montgomery in World War II. That would be to anticipate because all that was going on between statesmen and soldiers was being stored up in the mind of Hankey, even now often the successful intermediary, and was used by him to help to shape better machinery for the future.

Chapter 5

THE WESTERN FRONT: NEUVE CHAPELLE TO PASSCHENDAELE

A GUNNER'S WAR

Although in 1914 artillery had been considered the dominant supporting weapon, artillery preparation or bombardment was not a usual preliminary to the attack; the artillery waited to open fire until targets were disclosed by the infantry advance. The Battle of the Aisne showed that these tactics were not sufficient against a determined enemy in position; out of this battle came the lesson that artillery fire was no longer simply a useful adjunct to the attack, it was an essential element. Later, as the front became stabilized, the requirement of artillery became accentuated; for the first time the attacker faced the problem of breaking through an entrenched and wired position. Even in the mobile stages it became apparent that ammunition expenditure would be far greater than had been foreseen. The pre-war assumption was that 1,800 rounds per 18 pounder gun would be required for the first six months, after which an additional 500 rounds per gun would become available; but by the end of the Battle of the Aisne some guns had fired more than 2,000. In the first day of the Battle of Neuve Chapelle more field ammunition was fired than in the whole of the South African War. The shortage of ammunition had a great effect on the minds of those who were thinking out the problem of the attack, because it gave an easy explanation of the reason for failure. The idea that artillery was the real answer to entrenchments and machine-guns was not confined to the more orthodox soldiers. Colonel Swinton, who as much as any one man was responsible for the invention of the tank, was one of the first to identify the problem created by German strength in artillery, machine-guns, and wire. In his search for a ' power driven, bullet proof, armed engine, capable of destroying machine-guns, of crossing country and trenches, of breaking through

The Western Front: Neuve Chapelle to Passchendaele

entanglements, and of climbing earthworks', his mind turned one day in October 1914 to the Holt caterpillar tractor which he had seen at Antwerp just before the war. But as Swinton himself says, he thought artillery fire would do if we had enough high explosive.[1]

The Battle of Neuve Chapelle, which began on 10 March 1915, was the first British major offensive against an organized trench system. It was carefully thought out and thoroughly prepared. There was to be a sudden and intense bombardment to allow the infantry to advance as rapidly as possible on a 2,000 yards front and, before the enemy had recovered his equilibrium, to seize the Aubers Ridge one and a half miles beyond Neuve Chapelle. Guns would be brought into position gradually and secretly and registration of targets was to be as unobtrusive as possible. Special tests were carried out in rear areas to assess wire cutting capability and the length of bombardment necessary. It was found that shrapnel was the most effective wire cutter since, before the invention of the instantaneous fuze, H.E. churned up the wire but left a formidable obstacle. The preliminary bombardment was to last thirty-five minutes, which was considered sufficient to cut wire and destroy trenches.

Complete surprise was achieved and on the greater part of the front the first objectives were captured. As the attack proceeded artillery support proved insufficient, impetus was lost and the enemy was able to maintain his hold on Aubers Ridge. The result showed what could be done in the breaking of the enemy forward position, but it was too easily assumed that if there had been more ammunition a complete break through could have been achieved. In fact there were other reasons for the failure to maintain the initial success: the difficulties of communication between infantry and artillery and between commanders, because of the vulnerability of telephone lines under shell fire; the difficulty of getting reinforcements to the required place; and the bravery of a few of the enemy in holding on to gain time to restore the situation. Possibly, however, the greatest cause of the loss of momentum in the attack was the tendency to use reserves to repeat assaults that had failed rather than to exploit success.

[1] Swinton, *Eyewitness* (Hodder and Stoughton, 1932), pp. 79, 129.

A month later the Germans took up the offensive again at Ypres. This time they achieved technical surprise by the use of poison gas. A situation most dangerous to the Allies arose but the Germans had made no arrangements for reserves ready to exploit success, nor had they carried out any special tactical training for the troops following up the gas cloud. The attack petered out and the trial of the new weapon achieved nothing more lasting than to warn the Allies of the need for anti-gas measures. This operation, like the later appearance of the tank, shows how difficult it is to bring into use a new weapon on a large enough scale to achieve decisive results. A balance must be struck between the secrecy necessary to ensure surprise and the thought, explanation, and training needed to ensure efficient use.

The experience gained at Neuve Chapelle seemed to have given the British the key to success if only they could master the problem of exploitation. Hesitation after the first success had given the enemy five hours to recover. The technique of bringing artillery into action and of dealing with enemy targets stood the test of the whole war, and is, indeed, the basis of artillery tactics to-day. When, in response to a request by Joffre for a British attack in co-operation with a French offensive north of Arras, Haig was ordered to make another effort to capture the Aubers Ridge, he planned for a similar type operation. There was to be a forty minute bombardment to cut the wire, destroy trenches and strong points, and to produce a curtain of fire to prevent the move forward of enemy reinforcements. In order to avoid some of the difficulties which had arisen at Neuve Chapelle there were to be two converging attacks 6,000 yards apart; it was hoped thus to cut off some six or seven battalions and a number of guns, which would be dealt with by specially detailed units. There were also a number of tactical innovations. Specific batteries were detailed to be ready to follow up the advancing infantry so as to engage enemy targets in depth; batteries of mountain artillery and mortars were detailed as 'infantry artillery' to act in close support of brigades; not only were aircraft with wireless telegraphy provided for reconnaissance and artillery observation but three aircraft were specially standing by to order the capture of certain lines.

The Western Front: Neuve Chapelle to Passchendaele

However, the enemy had also studied the lessons of Neuve Chapelle and had learnt more about defence than the British had about attack. He now occupied, not a simple trench system, but a solidly constructed line of fortifications. The bombardment was a failure, large stretches of wire remained uncut and machine-gun posts intact. The attack which went in on 9 May, while the Ypres battle was still going on, was a failure, and after being pressed again on 10 May was broken off.

Aubers Ridge, for the time being, killed the concept of the short bombardment. The French, who now had abundant ammunition, had already come to the conclusion that at least four days' bombardment with a high proportion of heavy artillery was necessary to destroy enemy morale, break up his obstacle and strong points, and disorganize his communications. Attacking with these methods at the same time as Aubers Ridge, their Tenth Army had achieved considerable success. The British failure had, however, set free German reserves and Joffre pressed French to continue his First Army attack. Haig undertook to renew the offensive north of Festubert and this time he arranged a thirty-six hour bombardment to be followed by a night attack to gain a footing in the first two German lines and to exploit at first light. Rain and mist prevented the observation of fire considered necessary and in the event the bombardment lasted sixty hours. The first stages of the attack were successful and the enemy forced from a strong position he had taken some months to prepare. It seemed as if the new enemy position could be carried at once, but ammunition was short and fresh divisions not available, so the depleted and tired divisions were pressed into the attack again without adequate support. They responded gallantly but they failed. Thus grew up the impression that the British were on the right lines, that given sufficient artillery ammunition, and the necessary number of troops to persist in the attack, success would follow. They had learnt that meticulously prepared long and accurate bombardment could win initial success, but they had not yet realized that something more would be required for ultimate victory.

After the failure of the spring offensives, French, with the support of Kitchener, wished to put off any further attacks until 1916 by which time a strong British army with abundant ammunition would be ready. The ammunition position in 1915 can be judged by the fact that the Germans were producing 250,000 rounds of gun ammunition a day, the French 100,000 and the British only 22,000. The establishment of the Ministry of Munitions in May 1915 would take about a year to show significant results.[2] Joffre looked at the situation very differently. The French Army was at its peak in numbers and training while the whole country burned to remove the invader from so much of its native soil. Joffre was confident that provided the attack was made on a wide enough front he could break into the plain of Douai and seize vital German communications which were only fifteen to twenty miles from the front. The whole of the German army southward to Verdun would then be in jeopardy. His plan was to make a simultaneous attack in Artois and in Champagne converging on Namur. As the break was made cavalry and infantry in buses were to be ready to exploit success. For the attack in Artois Joffre required assistance immediately on the left of Foch's group of armies, that is between the built-up area of Lens-Loos and the La Bassée Canal.

The British did not want to attack at all but certainly did not want to do so in the Lens-La Bassée area. The ground was flat and open and flanked by strong defended localities in the built-up area. The British General Staff considered that to be successful their attack would have to take in a front of twenty-five miles and would require not less than twenty-six divisions supported by 1,150 heavy guns in addition to the divisional artillery. At this time, after allowing for the troops required for the remainder of the front, only nine divisions were available. French and Haig, who had personally reconnoitred the La Bassée area, considered that if the British must make a supporting attack in 1915 it should be another effort to get the Aubers Ridge. This should be accompanied by an attack on the Messines Ridge and so give the British a better line for the coming winter. Joffre was adamant; he believed that attacks so far from Foch's armies would be separate and divergent. He pointed out that all the experience of the war showed that the only means of preventing the enemy concentrating his artillery from both flanks was to attack on a wide continuous

[2] Edmonds, *op. cit.*, 1915, vol. II, p. 116.

The Western Front: Neuve Chapelle to Passchendaele 85

front. The argument was tilted in Joffre's favour by the German successes on the eastern front and the fears that unless something were done in the west Russia would make a separate peace. There was a strong feeling among her allies that Britain was too long in making a real effort in the war. Kitchener therefore instructed French that he was to co-operate with Joffre in a vigorous offensive.

Haig was entrusted with the plans for the Battle of Loos. He had available for the main attack six divisions, of which two were New Army divisions never before in action (9th and 15th, both Scottish), and the 3rd Cavalry Division as his reserve. French kept in his own hand the general reserve for the battle, three divisions, two of them New Army, and three cavalry divisions. The main problem which faced Haig was that of artillery since he had only enough to support a two divisional attack. He decided that the deficiency would be made good by the use of gas. Since the British had decided to retaliate after the German use of gas their special companies of the Royal Engineers had rapidly mastered the technique and Haig had been greatly impressed at a demonstration on 22 August. The use of gas raised serious problems because of its dependence on a favourable wind and because the French, who were not using gas, could not allow a variable date for the offensive. Another serious disadvantage of the gas was that it could be used only in the deliberate first phase so that for the subsequent phases the British would have inadequate artillery support.

Joffre insisted that the attack must go in on 25 September and French could not therefore give Haig any latitude for the date he was to begin his attack. Haig did, however, get agreement for up to three days' grace for his main attack. If conditions were not favourable for gas on 25 September he would attack on that day with two divisions only, waiting until the 26th or 27th for the main attack. In the early hours of the 25th the wind was from the south-west, which was favourable, but it was very light. Haig decided to launch the main attack and at 5.50 a.m. the gas was released. Forty-five minutes later the infantry advanced in a misty drizzle with almost no wind. Despite the difficulties caused by the gas hanging about in the damp calm all divisions except 2nd Division on the left succeeded in capturing the enemy first line and advancing beyond it. The 47th Division on the right secured a sound defensive flanking position facing the dangerous Lens area beyond which the French were about to attack.[3] But once again the enemy was given time to recover. Not only was Haig short of artillery for his subsequent attack but the reserves which he required to exploit success were not there. To make matters worse Foch's attack on the British right proved less successful than our own.

The command and role of the general reserve, particularly of the three divisions of XI Corps, had been the subject of some disagreement between French and Haig. Haig, who had had to use all six divisions of I and IV Corps in the first attack, wanted to have at least two divisions of the reserve under his own hand. Joffre was of the same mind and writing of the risk of holding the reserve too far back told French it was essential that they should be put at Haig's absolute disposal before the attack. French, however, insisted on keeping the reserve under his own hand and in his original orders placed them sixteen miles behind the start line for the attack. Haig protested and asked that the two leading divisions of XI Corps should be ready and rested not more than 4,500 yards behind the front line at the time the attack went in. French assured Haig that by daybreak on 25 September the two divisions would be where he wanted them.[4] French did in fact issue orders to this effect to XI Corps, but he delayed the move forward until after dark on the 24th. Owing to much crossing traffic, delays at the numerous level crossings, bad roads, and indifferent traffic control the divisions took the whole night for their march and the last parties were not in position until 6 a.m. on the 25th. So far from being ready and rested the new divisions were already tired and strung out when the attack began.

French was wise enough to want to be well forward in the battle, but he went forward without any pre-arranged signals plan. He was situated 20 miles away from his headquarters staff with only the French civil telephone for line communication and no line forward to his Armies. As soon as Haig had

---

[3] The French actually attacked at 12.45 p.m. some hours of daylight being required for observed artillery fire.
[4] *Ibid.*, pp. 275-6.

The Western Front: Neuve Chapelle to Passchendaele 87

reports of the success of his first assault he sent a staff officer to French to ask for XI Corps to be ready to advance at once. French waited two hours before ordering the advance and even then did not put the divisions under Haig. When Haig decided to put in XI Corps the German second line seemed to be breaking and the orders as received by 21st and 24th Divisions suggested that the enemy had been heavily defeated and that no organized resistance to their advance need be expected. By the time the two divisions crossed the original front lines it was dark and they were committed to a night advance over ground which they had never seen and in the face of a German defensive position which had been almost completely restored. Happily the orders for an attack on that position were called off but the two divisions spent a second night on the move and with no rest.

On 26th the attack on the German second line was met not, as Haig expected, by a shaken and disorganized enemy but by determined defenders well protected by deep and intact wire. After the failure of this attack the battle went on for seventeen days without any progress being made. German counter-attacks regained one or two positions but the line on 14 October was not very different from that reached by the original British assault on the first day. The French, who with their many guns and abundant ammunition supply had forfeited surprise, fared even less well. The Vimy Ridge, on which so much depended, remained in enemy hands. Ironically, one of the reasons given by the French for their failure was that reserves were pushed too far forward with the result that they suffered unnecessarily heavy casualties and there was less scope for their use where required.

Loos is of special significance because it was the first major battle in which the New Armies were committed. The 9th (Scottish) and 15th (Scottish) Divisions, part of 'The First Hundred Thousand', acquitted themselves as well as any troops in the battle. The 21st and 24th Divisions, part of the 'Third Hundred Thousand', did less well. It was certainly not that they lacked courage in the face of the enemy, for by the time they were launched to the attack their task was well nigh impossible. It was the staff work and the knowledge how to look after themselves that was lacking. The task that faced them and the physical effort required in the move forward was not unlike that which the original B.E.F. tackled so well in August 1914. Much can be attributed to the inexperience of some of the staff and of the regimental officers; no battalion had more than one regular officer in addition to the commanding officer who was, as often as not, a retired regular officer. Neither G.H.Q. nor the First Army staff can escape all responsibility because the roads forward were organized to cater for normal supply traffic rather than the through movement of battle formations.

French has been criticized for using raw formations for such an important role. It must, however, be remembered that by this time all the regular divisions were much watered down by the inexperienced drafts sent in to replace heavy casualties, this specially showing in the lack of experienced officers. French thought that divisions fresh from training at home might be more thrustful in the reserve role than depleted divisions accustomed to the more sluggish atmosphere of trench warfare. It is pertinent that the third division of XI Corps was the Guards Division with experienced commanders, staff, and battalions. Despite impeccable march discipline they fared little better in the move forward on 25 and 26 September, and although their fighting ability was unimpaired they could do no more when put into the battle than stabilize the position. The truth is that again the Germans had shown themselves one step ahead in evolving tactics to combat our offensive methods. They started the Battle of Loos with a fully organized second line of defence out of reach of any preliminary bombardment. It was only after the capture of the first line and the move forward of the artillery that a sufficiently strong attack could be mounted against the second line. This was a system of defence for which Haig's original plan did not allow and which he had insufficient artillery and ammunition to overcome. The lesson for the future was that it was a system proof against even the more deliberate and adequately supported methods used by Foch.

As a result of Loos French was replaced in command by Haig. French might have been the man to command a B.E.F. of six or so divisions, although he can scarcely be said to have risen to the occasion. He was certainly not the man to direct

The Western Front: Neuve Chapelle to Passchendaele 89

an army of thirty or more divisions and to work out the methods required to solve the problem of a totally new form of war. Haig had come out of the battle with some credit. Despite his objections to the place and nature of the attack to which he was committed, he had carried out his instructions in letter and in spirit. By the use of a novel weapon and by his bold decision to launch the attack despite doubtful weather conditions, he had surprised the enemy and so achieved almost complete success in the first phase. He was proved right about the placing and command of the reserves. Loos was a battle in which the size of forces was still such that the commander could exert considerable personal control. Here there were indications in the conduct of the battle which showed how much he and his senior commanders had still to learn. Haig's orders to 3rd Cavalry Division were given under a completely wrong impression of the situation. His subsequent orders to XI Corps were given without knowing that 3rd Cavalry Division had found the situation quite unsuitable for cavalry action. The failure to devise means of passing information on which commanders could give realistic orders was to prove one of the dominant causes of failure in the coming years. Another possible criticism is that Haig never really faced the problem of artillery support for the subsequent stages of the attack. Here Haig had two vindications: first, he had said before the battle that there was not enough artillery for more than a two divisional attack but had been overruled to conform with our allies; secondly that this problem of continuing artillery support was one which was most energetically tackled after Haig had become Commander-in-Chief.

From Loos onwards officers of the Royal Artillery took an increasing part in the tactical planning of the battles. Earlier there was no commander of all the artillery in any larger formation than the division, although at Corps, Army, and G.H.Q. there were senior artillery advisers. In the development of artillery tactics Brigadier-General J. F. N. Birch was beginning to have a dominant influence. He had been Commander Royal Artillery 7th Division at the battle of Aubers Ridge, and then at Loos all the artillery supporting I Corps had been grouped under him. When Rawlinson became Commander of the newly constituted Fourth Army, Birch became his artillery adviser and remained so during the whole of the planning for the Battle of the Somme, although in June he left to become Haig's artillery adviser at G.H.Q. Birch was instrumental in getting a proper artillery command system introduced so that each army and corps had its General Officer Commanding Royal Artillery. Under them were groupings of the heavy and medium artillery, particularly for counter-battery work[5] which was becoming increasingly important. The artillery thus got the command system which enabled it to fight what was intended to be an artillery battle with infantry in a secondary role.

Birch's influence on Rawlinson can be seen in the concept of the battle as a step by step advance, each stage covered by a meticulously prepared artillery fire plan. Haig on the other hand was still thinking of a quick rupture of the enemy line to let the cavalry through. In the discussions between Haig and Rawlinson during the months of preparation the difference in approach can be seen.[6] Haig was at this stage searching for the true strategy; he stressed the need for surprise, the need to take advantage of early success, and he studied the ground and saw which objectives once captured would give further opportunity. Rawlinson looked at the means at his disposal and the methods of attack. He planned a long bombardment on as wide a front as resources allowed and a direct advance all along that front. Haig was able to insist on certain modifications, such as the capture of the Montauban spur opposite the British right, but he was not able to overrule Rawlinson in essentials because he could not suggest anything to obviate the long bombardment necessary to cut wire.

The tactical concept of 1916 was that a sufficiently long and heavy bombardment would so destroy the defences and kill or demoralize the defenders that infantry following up would only have to take possession. Rawlinson had seen the difficulty of dealing with successive objectives in the earlier battles but now we had sufficient artillery and ammunition to undertake a succession of sieges. The barrage was to move in a succession of lifts and as soon as it lifted the infantry would advance. The

---

[5] That is fire directed against enemy batteries who were in a position to engage our infantry.
[6] For details of conferences and letters see *ibid.*, pp. 250-62.

The Western Front: Neuve Chapelle to Passchendaele 91

lifts were by rigid time-table which could not be altered or called back without several hours notice, nor at lower authority than corps headquarters. The rigidity of the system was made worse by the lack of understanding of the need for speed in the infantry crossing no man's land or of their keeping close to the barrage ready to assault. The infantryman was moreover so heavily laden that he could not well move at more than a steady walk. The idea of reinforcing the firing line for the moment of assault[7] still held good. In fact in a G.H.Q. instruction[8] it was stated that in capturing a trench ' a single line of men has usually failed, two lines have generally failed but sometimes succeeded, three lines have generally succeeded but sometimes failed, and four or more lines have usually succeeded.' The author probably had in mind the countryman's adage, ' One log can't burn, two won't burn, three may burn, and four logs make a fire.'

It is worth remembering that the Germans at Verdun and the French both used similar methods, although the Germans at Verdun made some use of patrols pushed forward ahead of the main assault. Haig did indeed suggest in June 1916 that the advance might be made by small detachments instead of waves. The three of his army commanders who were infantrymen, Plumer, Monro and Rawlinson, all opposed the idea and Haig as a cavalryman did not pursue it.

Joffre's idea for 1916 had been a Franco-British offensive astride the Somme. The British did not want the battle to begin before June, and preferably late July, so as to give time for the arrival and training of the New Army and the accumulation of ammunition. The Somme was not the choice of the British; again they would have preferred the Messines Ridge, but again Joffre wanted the attack where it could be closely combined with the French. As the toll of Verdun went on it became increasingly clear that the main, if not the whole burden of the attack must be borne by the British. Finally the decision was made that both British Fourth Army and the French Sixth Army would attack on 28 June. The attack was to be in full daylight. The French liked plenty of time for

---

[7] See Ch. 2, p. 39.
[8] Training of Divisions for Offensive Action, G.H.Q., May 1916. Reproduced *ibid.*, Appx 17.

observed artillery fire immediately before the attack and would have preferred 9 a.m. to the 7.30 a.m. that was fixed. The Allies had very detailed knowledge of the German defensive position; every trench had been photographed from the air. What we did not know was the depth and strength of the dug-outs. The ability of the Germans to remain safe deep down and come up in lifts with their machine-guns when the barrage passed on was a complete surprise. The only hope for the infantry thus lay in keeping right close up to the barrage.

The preliminary bombardment, planned to last five days, began on 24 June but as bad weather delayed the attack until 1 July it went on for seven days. It was followed by what was the embryo of the creeping barrage. Army orders drawn up under the influence of Birch, who just before the battle began had left Fourth Army to become Haig's chief artillery adviser, specified a barrage advancing 100 yards in two minutes, but details were left to be worked out by Corps. The five corps commanders differed somewhat in their application of the orders and divisional commanders had the last word in the fire plan. VIII Corps on the left specified fifty yards lift per minute; as here no man's land was wide the barrage which started on the enemy front line was much too far away to be effective cover for the advance. In most of the other corps too the barrage soon got away from the infantry, but on the right XIII Corps specified that the barrage should ' creep back by short lifts '. The orders went on:

> The lifts have been timed so as to allow the infantry plenty of time for the advance from one objective to the next, on the principle that it is preferable that the infantry should wait for the barrage to lift than the latter should lift prematurely and thus allow the enemy to man his parapets. The infantry should follow as close behind the barrage as safety permits.[9]

Opposite this corps no man's land was only 200 yards wide and the advance to the first trench was carried out comparatively easily. Subsequently, in stiff fighting both leading divisions of the corps, 18th and 30th, captured the whole of the corps objective on the first day. Great credit is due to the corps

[9] *Ibid.*, Appx 21, pp. 157-8.

commander, Lieutenant-General W. N. Congreve, and his staff, but special attention must be directed to Maxse,[10] commander of 18th Division, because he was an expert trainer who was to have an outstanding influence on the development of British infantry tactics. That he had some influence on the training that went on in XIII Corps cannot be doubted. All who knew him can testify to his dominating personality and to his ability to tell both superiors and subordinates how things should be done. Let this not suggest that he was dogmatic or pigheaded, he was above all a man who thought things out and who learnt from experience, other people's as well as his own.

Apart from XIII Corps only one division was completely successful on the first day. The 36th (Ulster) Division, inspired no doubt by the anniversary of the Battle of the Boyne, achieved the magnificent feat of capturing the whole of the Thiepval plateau. Alas, this was not used as an opportunity to exploit success but efforts were wasted in renewing attacks that had failed elsewhere, with the result that 36th Division remained unsupported. Assailed on both flanks the division suffered heavily and the Germans regained most of the plateau. For four out of the five corps the attack was thus a failure. Casualties were heavy; at 10 p.m. Rawlinson, in touch with all his corps commanders by telephone, estimated them at 16,000. They were in fact 60,000. The French on the right had succeeded in capturing all their objectives and Rawlinson still had hopes that his second attempt would succeed.

The first phase went on for more than a week. By 8 July no ground had been gained in the two left corps, but the other three were in possession of most of the German first line. Rawlinson made his plans for the attack on the second line and in deference to the whole body of infantry opinion decided that this time the attack should be delivered at dawn so as to give some chance of defeating enemy machine-guns. The attack would be made on 14 July by two divisions each from XIII & XV Corps supported on the left by a subsidiary attack by X Corps. Haig had considerable doubts whether the training, discipline, and staff work were sufficiently good to allow the night move forward of half a mile and the forming up, followed by an attack in the right direction at dawn. The French thought

[10] See Ch. 2, p. 43.

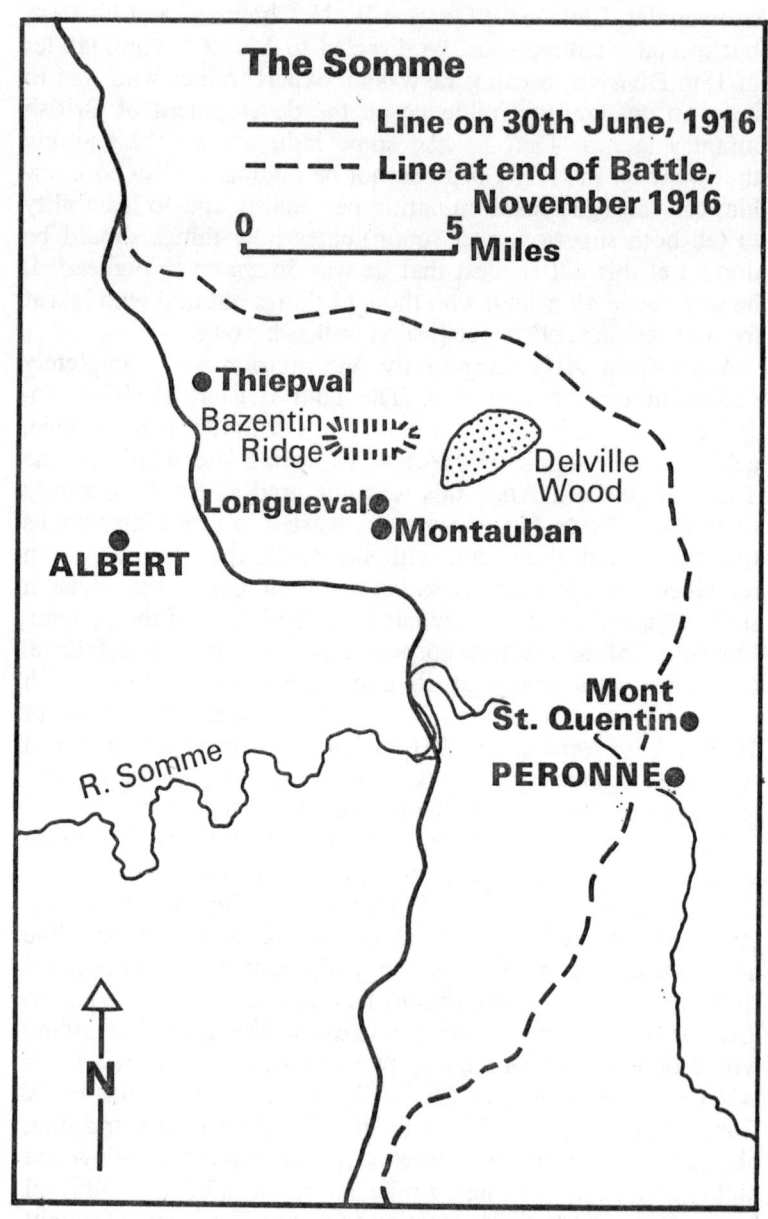

Map 4.

The Western Front: Neuve Chapelle to Passchendaele 95

the British were mad to try so complicated an operation with inexperienced troops, even on a night with the moon almost at the full. Rawlinson, supported by his corps and divisional commanders, was confident that his plan was practicable and pressed the matter. This time it was Rawlinson and not Haig that was determined to stake all on surprise, and Haig gave his consent. Brigadier H. H. Tudor, Commander Royal Artillery of the 9th (Scottish) one of the attacking divisions, was a moving spirit in devising a new type of artillery plan. He was against a preliminary bombardment to alert the enemy before the assault. He preferred to do the wire cutting in what would appear a haphazard programme over a few days and to hold the intense fire until the moment of the infantry assault. The invention of the new delay action fuse made possible a true creeping barrage using H.E. to which the infantry could safely keep close. The operation was almost completely successful and 6,000 yards of the second position were quickly captured. The Germans testify to the surprise and to the fact that the assaulting divisions took advantage of initial penetration and rolled up the defences to right and left. The 3rd and 7th Divisions were ready by 10 a.m. to exploit but were told to leave that to the cavalry and to prepare to meet counter-attack. The cavalry were not in position to move to the attack until 7 p.m. by which time it was too late.

The success of the dawn attack encouraged Rawlinson to persevere with night operations and he next planned for a night attack on 22/23 July. The artillery lesson had, however, been forgotten and this time the bombardment started the night before. The enemy was alert and the attack which was meant to capture Bazentin Ridge, Delville Wood, and Longueval and then go on to take High Wood and the whole of the German switch line gained no ground at all.

Despite the failures and disappointments Haig determined to go on and to destroy the German Army. The battle was not called off until 12 November, and before this, on 15 September tanks were used for the first time. Their advent was a complete surprise to the enemy, but they were not used, nor indeed available, in sufficient numbers to achieve decisive results. Like the Germans with gas, we used a new device before we were ready to reap the full benefit it could have given. Whether this was a mistake is a subject on which there will never be agreement. The earlier mechanical troubles of the tank might well have been eradicated by longer preparation and production but it is doubtful if the evolution of suitable tactics for infantry and artillery working with tanks would have been possible without battle experience. Certain it is that tanks used on the Somme were used in an almost diametrically opposite manner to that which the true students of the tank advocated. Both Swinton and Fuller advocated that the real purpose of the tanks was to accomplish the task, so far impossible to infantry and artillery, of making a quick breach through the whole depth of the enemy position. For this, large numbers of tanks on a wide front were necessary. The question is discussed in more detail in the next chapter. As a result of our premature use of tanks some German commanders did study the use of artillery against tanks with serious consequences for the Allies in 1917 on the Chemins des Dames and at Cambrai. On the whole, however, the Germans did not make good use of our revelation; they certainly did not, as we had done with gas, take the opportunity to develop their own weapons.

The later operations on the Somme were little more than a wearing down process of the Germans and a continuing relief to the French at Verdun. The chief lesson is the self-defeating effect of concentrated artillery fire, a lesson which, unhappily, was not learnt. In the rain which persisted through most of October it was almost impossible for the heavily laden infantry to move even without enemy opposition. It is, however, worth studying for a moment the performance of Maxse's 18th Division to see the development of his tactical thought.[11] After the dawn attack on 14 July, 18th Division had gone off to Plumer's Second Army in Flanders. In September it returned to the Somme to join Gough's Fifth Army which had been formed out of the left two corps of Fourth Army. On the way back the division had carried out three weeks' intensive battle training. Their first task was to capture Thiepval and the Ridge which had been taken and lost by 36th (Ulster) Division on the first day of the Somme Battle. Maxse's training bore good fruit. The leading battalions were out of their assembly trenches before zero hour and were close up to the first objective by the

[11] This and subsequent references to Maxse are based on the Maxse Papers.

Plate 1. H.M. King George V talking to a private at an Infantry demonstration in preparation for the Third Battle of Ypres. Note the load carried by the soldier.

Plate 2. A tank crashing through barbed wire at Wailly, 21 October 1917.

Plate 3. Part of the barbed wire entanglement in front of the Hindenburg line near Heninel, at the Battle of Arras, 3 May 1917. It can be seen that artillery fire made little effective impression on such deeply set wiring.

The Western Front: Neuve Chapelle to Passchendaele 97

time the barrage lifted. The supporting battalions too were clear of the front line when the inevitable enemy bombardment came down there. The capture of a place like Thiepval Ridge was not to be achieved without the bitterest struggle, but eventually all objectives were taken. Maxse's forethought and organization, his teaching on trench clearance, mopping up, consolidation, and above all on the close liaison between artillery and infantry all paid a handsome dividend. In the German counter-attacks which followed, strong points were lost and regained, but in the end all objectives were held except the western face of Schwaben Redoubt. Haig paid a special visit to the division to congratulate them and Plumer wrote to Maxse, '. . . send me a few lines describing your attack and how you carried it out—as you say trouble in training and careful organization of the smaller units pays over and over again.' It is significant that after the battle Maxse studied in great detail all that his brigades had done, where their methods had succeeded and where they had failed. He also carefully analysed the casualties and all was thought over for future operations.

In the last phase of the Somme, the battle to capture the salient north of the Ancre, 18th Division were equally successful. In his report on the battle Maxse said that the secret of successful attacks in modern trench warfare might be summed up in the words 'Previous Preparation'. He summed up the doctrine of 18th Division as follows:

> Teach, drill, and practise a definite form of attack and see that every man knows it thoroughly. On this basis of theory and knowledge common to all, any brigade, battalion, or company commander varies his attack to suit any condition which may be peculiar to his front and his objectives . . . [There is] neither the time nor the opportunity to learn a new attack formation before the assault. They can, however, invent, explain, and vary . . . as they did at Thiepval.

Maxse was much against the practice of the attack at first light. He believed the time of attack should be worked backward from sunset. As a rough guide he wanted two or three hours of daylight on the final objective to start the work of consolidation and to allow work through the night clear of observed fire.

Although Maxse was insistent that tactical training should be approached as a drill, he was not in favour of a stereotyped form of attack. During the Somme he wrote:

> I rather feel we are trying to do every attack by barrage. The method of applying it is identical and I feel certain the Bosche knows we are going to attack where and when we are. I suggest where the wire is cut it would be better to attack without a preliminary intense bombardment and only lay on a barrage after the attack has begun, and then only to stop reinforcements being sent to the position attacked.

Even from the days of training at home he had stressed the necessity for close liaison between infantry and artillery right down the chain of command. He insisted on battery commanders being present when infantry brigadiers gave out orders, and on artillery officers being forward with infantry battalion commanders.

Maxse was critical of orders such as ' to take place immediately ' or ' capture at all costs '. These were a legacy of the days when the assault was the culmination of a manoeuvre in the open. Now the situation was different. Maxse knew that the battalion commander must have time to see the ground and to form up his battalion square to the objective. He said that company commanders who sent detachments round the flank were more successful than those who charged straight ahead. He demanded initiative: 'Although there are few opportunities for initiative there are opportunities when initiative is essential for success . . . If we could devise more means for training subalterns to take such initiative, we should probably incur fewer losses and inflict more on the enemy.' Above all Maxse emphasized the need for training and for every task in or out of battle being done by men in their own sub-unit under its own commander. He envied the French system by which every man before he joined his unit in the line went through a definite course of instruction under a battle experienced instructor. That not every British commander regarded field training in the same light is illustrated by an acquaintance of the author who in 1917 joined his battalion of the Brigade of Guards. It was a

The Western Front: Neuve Chapelle to Passchendaele 99

fine battalion and the commanding officer, a magnificent fighting soldier, greeted his young officer straight out from Sandhurst with the words, ' You can forget all your training; you have come out here to show your men how to die.'[12]

It is difficult to strike a balance of advantage in the Battle of the Somme. The British Army that went into action on 1 July was a large slice of the finest manhood of the nation. Britain had never put an army like it into the field. It is not unfair to say that the army was regarded as of amateur status, expected to win its battle by dash and bravery rather than by any tactical skill. Some of the higher commanders had by this time much experience of war but on the whole they were raw in the handling of large bodies of troops. Only two of the corps commanders had commanded as much as a division and only three of the twenty-eight divisional commanders as much as a brigade in peace or large scale exercises. The very bravery of the regimental officers and the dauntless courage of the men who had followed them had led to unnecessary casualties. So often it had seemed that one more effort by these magnificent men would turn the scale and too often the higher commander failed to see that he was asking the impossible or that the objective could be achieved in a better way. With the Germans it was very different. At the beginning of the battle they were still a highly trained professional army commanded by men who had studied war on a continental scale all their lives and who had much experience in handling large formations in training and in war. Throughout the battle the Germans studied the tactical problem with which this appalling new type of warfare faced them. It must be confessed that in their understanding of the basic principles of fire and of movement they were always ahead of the British. And yet in the whole of the battle their casualties, totalling just over 650,000,[13] slightly exceeded those of the British and French combined. The German Army which came out of what they called ' The Hell of the Somme ' was never again the same fine instrument. On

---
[12] Letter to the author, 30 Aug. 1965.
[13] Edmonds, *op. cit.*, 1917, vol. I, p. ix. Liddell Hart who has made a special study of the subject considers Edmonds grossly exaggerates the German losses. Cyril Falls, however, writes, ' It is just possible that Edmonds exaggerates slightly, but I prefer his figures to those of Liddell Hart.' (Letter to the author, 18 Feb. 1968.)

the other hand, despite what it had suffered, and, whatever the impression conveyed in the anguished yet beautiful writings of some of the war poets,[14] the British emerged with its spirit unimpaired and ready for the awful burden which it was to bear in 1917.

We have already discussed the respite which the fighting on the Somme gave to the French at Verdun and the supersession of Joffre by Nivelle.[15] The British part in the Nivelle offensive was an attack by Allenby's Third Army in the Arras area directed on Cambrai. The left flank of this attack was to be protected by an attack by First Army to seize the Vimy Ridge. This ridge was one of the main bastions of the German defence line; its importance can be judged by the efforts the French had made to capture it in the spring and autumn of 1915, efforts which had cost them 150,000 casualties but had done no more than narrow the German hold on the ridge. Despite the cramped nature of the German defences they were made strong by elaborate redoubts and machine-gun emplacements, and a garrison protected by numerous deep dug-outs and natural caves. Although the concentration and forming up of the Canadian Corps was aided by a magnificent system of tunnels and subways constructed by the Royal Engineers, the chief surprise to the Germans when the attack came in on 9 April was the thoroughness of the counter-battery fire and the skill and dash of the Canadian infantry.

After the battle, Byng, commander of the Canadian Corps, writing to his friend and brother cavalryman Chetwode, then in Palestine, said:

> My own corps took the Vimy [sic] with a bang on 9 April doing every hole in bogey, picking up over 5,000 prisoners and 70 guns . . . They knew we were coming but were demoralized. You see we began the battle towards the end of February and kept a steady bombardment on the O.Ps day and night, stopping the rations, reinforcements, relief . . . In places they fought well but we had rehearsed taking the strong points so they did not come as a surprise.

[14] Strangely enough some of the best and probably the most soul-searing were written by three officers who served together in a very fine fighting battalion; Graves, Sassoon, and Owen. 2nd R.W.F.
[15] Ch. 4, pp. 70-71.

The Western Front: Neuve Chapelle to Passchendaele

Going on to talk of the main battle he wrote: 'The Bull's [Allenby] show ran equally well and it was a real licking for the Bosche and he cannot hide it. Great expectations were centred on the French push and they talked about breaking through (we gave up that catch word long ago) and " the line of the Marne ".'[16]

Vimy Ridge was only an accessory to Allenby's attack and though Byng deprecated the use of the word ' breakthrough ', the attack at Arras, better managed might have achieved just that. In the planning stages there had been the same controversy as over the Somme. This time it was Haig who urged the long bombardment and Allenby who wanted a short hurricane bombardment to facilitate surprise. It was a gunner's argument and Birch advised Haig that the forty-eight hour hurricane bombardment, advocated by Holland, Major-General Royal Artillery of Third Army, could not cut the wire and would not so severely try the enemy's morale as a longer bombardment. Haig's Chief of General Staff wrote to Allenby:

> As a result of past experience it may be said definitely that in view of the great and prolonged preparations required, the enemy cannot be surprised as to the general front of an attack on a large scale, but only to some extent as to its exact limits and as to the moment of assault.[17]

As a result of their careful study of the battles of 1916 the Germans had devised new defensive tactics. They realized that the linear forward defences were never capable of holding off a prepared attack, and involved unnecessary losses. They substituted for them an outpost zone of fortified posts and machine-gun nests in great depth. Action by small garrisons would thus give time for large scale counter-attacks, for which divisions were retained behind the second line of defence. Much of the artillery was kept out of range of the counter-battery programme but sufficiently far forward to shell the attackers when they had broken into the outpost line and were the subject of counter-attack. Some idea of the new system had been given in a translation of a captured document produced by the British

[16] The Chetwode Papers, Imperial War Museum.
[17] Edmonds, *op. cit.*, 1917, vol. I, p. 178.

Intelligence Service in February 1917, but no great study was made of it.

The Hindenburg Line began south of Arras so only the right of Third Army front was affected by the German withdrawal in March. Third Army attacked on a three corps front and the cavalry was in reserve on the right ready to exploit south of the River Scarpe towards Cambrai. When the attack went in on 9 April it was XVII Corps on the left which was successful. The 9th (Scottish) Division captured all three of its objectives, despite the fact that the heavy wire protecting the third line was untouched by the bombardment. One of the reasons for the success of 9th Division was the carefully laid on barrage under the direction of Tudor, the C.R.A. No shrapnel was used in the barrage, it was used only in the protective curtain on the objectives. In the high explosive barrage one shell in four was smoke. The barrage conformed exactly to the undulating ground over which the advance took place, with the result that the infantry were able to keep close up to it. One of the commanding officers, Croft, reported to his Brigade Commander that there was an opportunity for the cavalry ' but it must be done now, to-morrow will be too late '.[18] The division was ready to go on but was prevented by protective bombardment and then by orders that 4th Division should pass through as arranged. Carton de Wiart, a Brigadier in 4th Division, spoke to a cavalry liaison officer in exactly the same terms as Croft. These two divisions had taken part in an advance of three and a half miles, the largest advance on the western front made in a single day since the onset of trench warfare. But the opportunity was not taken. The cavalry were waiting south of the Scarpe for an opportunity which never occurred. The command system was not flexible enough to take advantage of success when it came. Worse, when the opportunity had passed the attack was pressed beyond hope of success. Byng in the same letter already quoted wrote of the cavalry in action against Monchy le Preux, a feature giving a long wide sweep over the whole Douai plain:

> The poor old cavalry got it in the neck at Monchy . . . they did splendidly and hung on to Monchy like heroes. The 10th

[18] *Ibid.*, p. 237.

The Western Front: Neuve Chapelle to Passchendaele

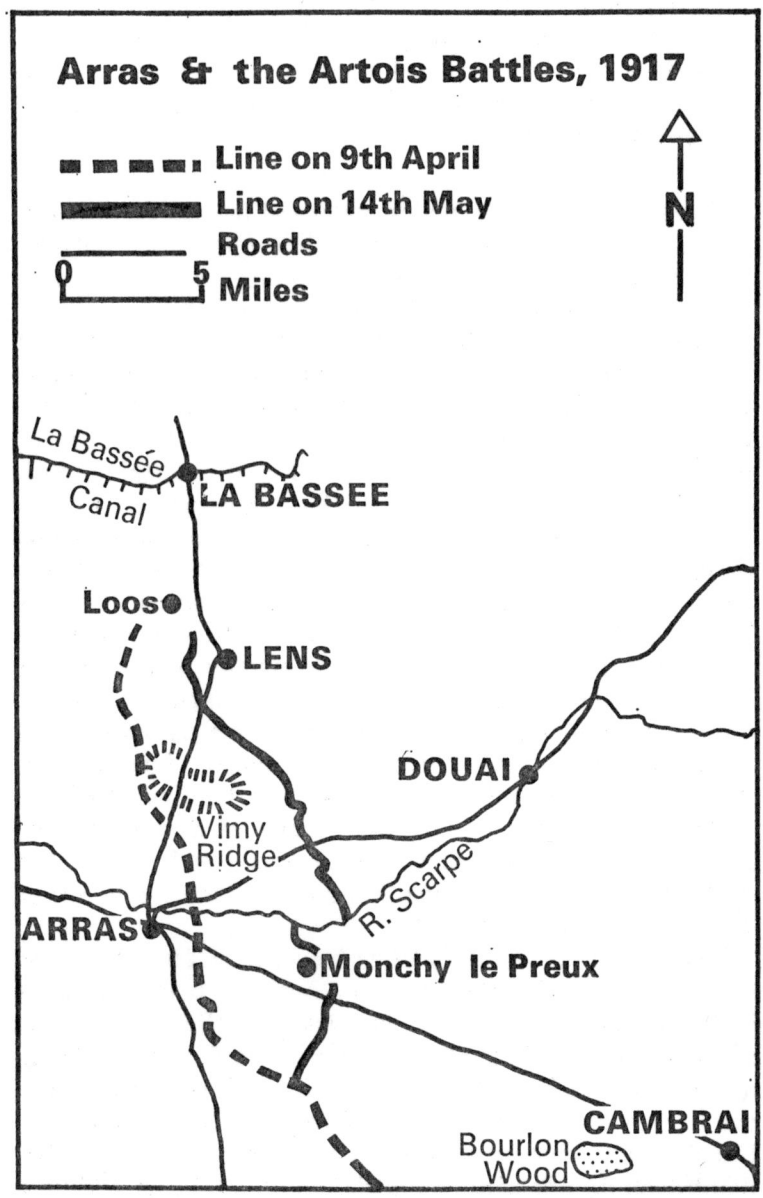

Map 5.

[Hussars] lost 8 officers, 220 other ranks and 350 horses and the Essex Yeomanry about the same. They were only in action about three days and it seems a pity to lose all these chaps who were perfect cavalry for a village which is a shell trap for either side.

This was Allenby's first major command in a set-piece battle, and a month later he was succeeded in command by Byng. He went on to command in Palestine where he showed how much he could achieve in the more fluid conditions there.

Great skill and dash had been shown by the Canadians at Vimy and by XVII Corps north of the Scarpe but on the whole the lesson of the Battle of Arras was lack of flexibility. Senior commanders had not yet devised means of keeping abreast of the situation. Staff officers were not used to keeping touch with forward battalions and brigades. Battalion and brigade commanders were tied down to command posts and it was thought more necessary that a commander should be where he could be reached by the superior headquarters than that he should be well forward where he could see what was going on. The result was not only the lack, already pointed out, of ability to take opportunities but that commanders did not know the situation in which their orders were received. Battalions were hustled into battle without proper time to study their problem and issue orders. Yet when time was vital the sense of urgency was often lacking. Some of these tendencies had already been commented on by Maxse and a few other divisional commanders during the Somme battles but the lesson had not yet fully been learned.

The failure of Nivelle's offensive coming on top of the appalling losses at Verdun led to a series of serious mutinies in the French Army. Pétain succeeded Nivelle in command and Foch became Chief of Staff in Paris. The French weakness and disorganization, although it was a well-kept secret, put the burden of the western front firmly on British shoulders. For this and other reasons which have been discussed, Haig decided on an offensive in Flanders. The attack by Plumer's Second Army at Messines Ridge from 7 to 14 June showed once again what could be achieved when the attack was

The Western Front: Neuve Chapelle to Passchendaele

thoroughly prepared and the objective limited. It was on a much larger scale than the capture of Vimy Ridge and carried out in the same skilful manner. The success might have been exploited to form a useful jumping off ground for an attack on the Passchendaele Ridge but Gough, whose Fifth Army was to carry out the main offensive, preferred to wait for an attack along his whole front.

Haig felt that his army commanders had failed to take advantage of an outstanding success at Messines, but as he realized, he could not be absolved from all blame himself. Although he had long contemplated action in the northern sector his attitude to the operation as planning developed and as circumstances changed was equivocal. His first plan was made in December 1916 and gave the task of working out the details to Second Army. Plumer, who knew every inch of the ground, recommended a stage by stage advance with one army capturing Messines and another striking north-east from the Ypres salient. Rawlinson was then brought into the planning. As Nivelle's optimistic ideas of lightning blows gained political acceptance Haig's mood changed. He was sceptical of Nivelle's views but nevertheless believed that since the British Army was to be relieved of the leading role in the spring offensive they might be able to deliver a second great blow which would clear the whole of the Belgian coast. He thought that, 'The plan should be based on rapid action and entail the breaking through of the enemy's defences on a wide front without any delay.'[19] Even after the failure of Nivelle's offensive and the stalemate at Arras that Haig still believed in a rapid operation of great scope is shown by his decision to put Gough in command of the northern attack. Gough, aged forty-seven, was the youngest of the army commanders and had first been appointed to command an army to take advantage of the expected success of the Somme battle and to lead the exploiting forces through the gap the Fourth Army was to make. However, as the effects of the Nivelle offensive began to be apparent, Haig realised that the character of the northern offensive must change. In a memorandum to the War Cabinet on 1 May he wrote of his plan:

[19] Letter from G.H.Q. to Plumer, 6 Jan. 1917, reproduced *Ibid.*, Appx V.

We shall be attacking the enemy on a front where he cannot refuse to fight and where, therefore, our purpose of wearing him down can be given effect to, for the first step must always be to wear down the enemy's power of resistance until he is so weakened that he will be unable to withstand a decisive blow.[20]

This new feeling was apparent in the G.H.Q. instruction which was handed to Gough as a basis for his planning. The Messines operation, as we have seen, was carried out independently by Second Army. For the northern operation the first objective was to be limited to the capture of the German second line, an advance of about a mile, then there was to be a pause of two days to move the artillery forward and to complete the securing of the Gheluvelt Plateau. Throughout his talks and correspondence with Gough, Haig stressed that the main battle would be fought on the high ground west of Gheluvelt and that Gough must plan accordingly. On the other hand Gough said in 1944:

I have a very clear and distinct recollection of Haig's personal explanations to me, and his instructions, when I was appointed to undertake the operation. He quite clearly told me that his plan was to capture Passchendaele Ridge, and to advance as rapidly as possible on Roulers. I was then to advance on Ostend. This was very definitely viewing the battle as an attempt to break through, and moreover Haig never altered this opinion till the attack was launched, so far as I know.[21]

On the day the Messines operation ended Haig described the operation as an offensive to 'secure the Passchendaele-Staden-Clerken Ridge'. Fourteen days later at a conference of his army commanders, Haig said to Gough, 'The object of the Fifth Army offensive is to wear down the enemy, but at the same time, to have an objective. I have given two: the Passchendaele-Staden-Ridge and the coast'. To Plumer he said:

Be ready to act offensively on the right, north of the Lys. I shall then have three fronts; one facing north east; the

[20] Edmonds, *op. cit.*, 1917, vol. II. p. 21.
[21] Quoted *Ibid.*, p. 127.

The Western Front: Neuve Chapelle to Passchendaele 107

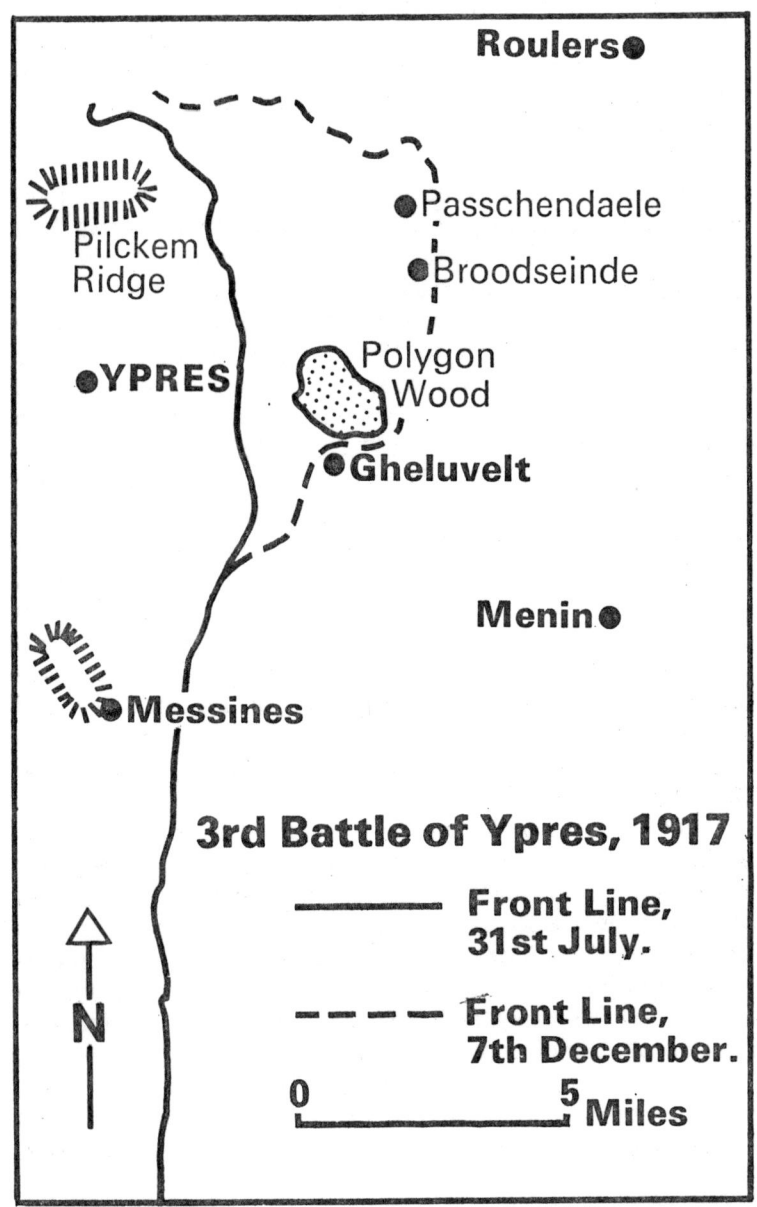

Map 6.

centre facing east; and the right facing south east. In view of the great possibilities accruing from the operations, I wish to be able to operate offensively at will from any of these fronts according to the way the enemy disposes his forces to meet me.[22]

The attack finally went in on 31 July. The preliminary bombardment began on 16 July and the air offensive five days earlier. Never had the Germans been more aware of the general area and time of an offensive. Some advantage came to the Allies from the German certainty because they moved every available man and gun to Flanders and so failed to take advantage of the French weakness. This fortuitous benefit could, however, hardly be claimed to excuse the failure to achieve any degree of surprise. Nevertheless the attack on 31 July attained a considerable measure of success, far greater than on the Somme and at less—though still grievous—cost.[23] By nightfall almost the whole of the German second line was in Allied hands, and on the left they had gone well beyond it, capturing the whole Pilckem Ridge. Despite violent German counter-attacks on both flanks, most of the captured positions were maintained. On the other hand the new line was only half-way to the final objective fixed for the first day and although they had lost much of the original artillery observation line, the Germans still maintained a strong hold on the Gheluvelt Plateau.

The Germans had by this time perfected the new defensive technique which had been in its early stages, and not altogether successful, at the Battle of Arras. They no longer used lines of trenches in the forward areas but a system of strong-points and the use of machine-guns in shell holes. Fear of our counter-battery fire had forced the guns further back, but still able to support the counter-attack effectively, and still making our attack on distant objectives difficult.

Rain began on the evening of 31 July and continued incessantly for three days and nights. The attack had been supported by 117 tanks of which seventy-seven were put out of action

---

[22] *Ibid.*, p. 131.
[23] Casualties for the three days 31 July-2 Aug. were 31,850 against 57,540 on the first day of the Somme where less troops were engaged.

The Western Front: Neuve Chapelle to Passchendaele

(forty-two were a total loss) either by the enemy or by ditching or mechanical breakdown. Until the weather improved tanks were not, therefore, available, but the Fifth Army continued its attack throughout one of the wettest Augusts in recent times, without appreciably altering the position. Gough made the mistake of using his artillery resources evenly along the whole front, whereas the Germans concentrated their artillery, both defensive fire and counter-battery, on action designed to maintain their hold on the Gheluvelt Plateau.

At the end of August Haig decided to hand over the main weight of operations to Plumer, and all corps of Fifth Army, except the two northernmost, came under command Second Army. This change confirms the opinion that Haig would have done better to have left the leading role in Plumer's hands throughout and to have made the Messines attack part of the main operation. As it was the change in command lost three weeks, time which was all the more important because September was a dry month.

In the meantime there had been an important limited success further south where the First Army had captured Hill 70, near Lens. The accurate well prepared artillery programme combined with the skill and dash of the Canadians was irresistible.

The offensive was renewed with a series of carefully prepared step by step advances on 20 September. The artillery was used to destroy obstacles and strong points and to isolate and demoralize the counter-attack garrisons and gun crews. An effort was made to organize the artillery fire so that it would not create new obstacles to the Allied advance. Their infantry tactics were beginning to loosen up; the main assault still went in in line to follow the creeping barrage but skirmishers were used, and there were dispersed groups ready to manoeuvre against strong points; special mopping up parties were also detailed. Second Army Instructions laid down that every unit down to platoons should keep a quarter of the unit as a reserve. The result of the battle (Menin Road) was that Second Army succeeded in establishing control of almost the whole of the Gheluvelt Spur, one strong enemy locality known as Tower Hamlets being largely missed in the barrage and holding out. The Fifth Army on the left also made some advance. The attack was resumed in the Battle of Polygon

Wood on 26 September. The objectives were gained and held, including Tower Hamlets, but this time our casualties were much more severe, both infantry and artillery suffering heavily.

These two successes determined Haig to go on, despite the lateness of the season. He hoped at any rate to capture the Passchendaele-Staden Ridge by the end of October. On 4 October the rain began again in earnest and it is this month which gave to the Third Battle of Ypres its notoriety and the name Passchendaele. Despite the rain the first action at Broodseinde was a considerable success, Second Army capturing more than 100 officers and 4,000 men in one day. Haig was convinced that two more blows would get the cavalry through. What Haig did not realize, and what the divisions and the corps fighting the battle should have been able to tell him, was that there were still in front of him two formidable positions, protected by unbroken belts of wire forty yards deep. In the First Battle of Passchendaele practically nothing was gained, in the Second, the Canadian Corps, as at Vimy and Hill 70, fought magnificently to capture the high ground at Passchendaele, but the battle ended without either Second or Fifth Army getting a hold on the main ridge.

During the Third Battle of Ypres Maxse commanded XVIII Corps which throughout formed part of Fifth Army. His notes show that he was constantly stressing the need for junior leadership and for training. He believed that even in this kind of war ' flank attacks are possible. Change of direction involved is not difficult. Our barrage is good guide. When a few men get behind the Germans they are willing to surrender.' He also stressed that the rifle was coming into its own again. Perhaps the most significant note, showing the way his mind was working was:

> The customary system of attacking in a series of waves has been found by experience to be unsuited against pill box defence. Now a definite pill box or machine-gun emplacement must be allotted to each unit. Unit behind barrage in column, protected by skirmishers. In most cases two sections or a platoon suffices as a unit.[24]

---

[24] Maxse's notes on the Battle of Menin Road, 20 September 1917.

Maxse's theories were based partly on his own experience and partly on a careful study of a captured German document on their new methods in defence. In his comments on this document, which he sent to Fifth Army, Maxse put forward valuable ideas on the use of tanks, low flying aircraft, and infantry in the attack.[25] He stressed the need for a more extensive use of low-flying aircraft against strong points, machine-guns, and infantry massing for the counter-attack. He wanted the infantry formation for the attack to be ' built up of platoons working in depth rather than battalions stereotyped in waves '. He also recommended the introduction of mechanical carriers to move with the infantry to lighten the load on the soldier. His paper went on:

The requisites for the effective use of tanks are
   i) Sufficient aeroplanes to deal with enemy anti-tank guns and forward batteries.
   ii) Area to be traversed should not be made impassable by our own heavy guns during the preliminary bombardment.
   iii) Smoke to blind enemy artillery observation posts.

Tanks should be used in sufficient numbers to neutralize strong points, nests of machine-guns and generally to facilitate the tasks of the infantry. Tanks should be given definite objectives and definite infantry units should be told off to co-operate in attacking each tank objective, but not necessarily to accompany the tanks. A suitable formation for tanks would be in several echelons according to the depth of the objective. The infantry formation (which would have been already practised) aims at stalking the tank objective under a barrage and assaulting simultaneously with the tanks.

On 21 August 1917 he wrote also expanding his views on the infantry attack:

The leading infantry, following the creeping barrage, should be in wave formation. Behind this leading infantry we require " worms " not " waves ". Instead of crowds of men in rear of the front line we want little columns of units in depth. We want each leading platoon to stalk a particular

[25] Quotations from the Maxse Papers.

farm or hedge or area of ground and to stalk it on a narrow front close to the barrage. In fact, we must have more power of manoeuvre than can be got out of the wave formation for supporting troops.

These two papers show how far ahead of his time Maxse was in his tactical thinking. Neither of them would have been out of place at any time in the Second World War. We see the first sign of Maxse's revolt against the advance in line, a formation particularly vulnerable to machine-gun fire from a flank.

Third Ypres was the last of the battles of attrition in which the sole recipe for success was to blast a hole through the enemy lines with artillery and engineer explosives. Both sides had tried the same methods and neither had ever achieved more than limited success. In the fighting from 31 July to 12 November the Second and Fifth Armies together suffered 238,313 casualties. Of this almost half were after 3 October. The German losses have never been disclosed. General Edmonds, the British official historian produces evidence to suggest they were in the nature of 400,000.[26] The battle was a brilliant feat of arms only in the sense of the magnificent courage shown by those who took part. In retrospect, perhaps, a justification for the offensive was the necessity to cover the French weakness; that this end was achieved without breaking the spirit of the British Army is a memorial to Haig and to those who fought under him. The evil reputation of the battle was in part deserved, but only in part. Much of what has been written and thought about it was inspired by Lloyd George's brilliant tongue and by his every effort to discredit Haig and Robertson.

[26] Edmonds, *op. cit.*, 1917, vol. II, p. 363. Again, Liddell Hart considers this a gross over-estimate.

Plate 4. The wire of the Hindenburg line penetrated by the Australians and other troops near Bony, October 1918. Note the depth of the wire and how the lay out of lanes and oblique sections facilitate defence by machine guns from a flank.

Plate 5. Infantry and tanks waiting for the second stage of the attack near Warfusee-Albancourt, August 1918. Note the infantry is drawn up in a formation suitable for advancing in depth rather than line and for making the best use of ground.

Plate 6. Canadian Mounted Rifles in training with tanks at Bony, 12 June 1918. The 'worm' formation advocated by Maxse had not been adopted here.

Plate 7. German barbed wire defences of the Hindenburg line at Queant, which was captured on 3 September 1918.

Chapter 6

CAMBRAI

THE EVOLUTION OF TANK TACTICS

The fighting at Passchendaele had hardly died down before Haig struck one more blow. This time it was on dry downland south of Cambrai and a very different type of battle. The Battle of Cambrai, which achieved so much at first and then seemed to go to nothing, has been the subject of much misconception and misunderstanding. Yet it did mark a new phase in warfare and the study of its lessons by a few thoughtful soldiers led to the development of a technique which at last made it possible to reap the fruits of Haig's understanding of the strategy of the offensive battle.

Cambrai is remembered as the battle in which tanks were for the first time used properly. Its real military importance is that it is the meeting point of two schools of thought—the apostles of the tank and the architects of the artillery fire plan. The early stages of Third Ypres had shown once more that tanks could not operate effectively in heavily shelled wet ground, nor could they keep pace with the infantry over ground churned by the preliminary bombardment. As a result Brigadier-General Elles, commander of the Tank Corps, had conceived a plan for a large scale raid on the Third Army front. There was to be no preliminary bombardment; support was to be provided by low flying aircraft. As the plan was developed at Third Army headquarters it was decided that effective artillery support could be provided without a preliminary bombardment and without forfeiting surprise.

Before examining events further we must trace for a moment the thought and the developments which made the Cambrai plan possible. Reference has already been made to the birth of an idea in Swinton's mind in 1914 and to its development in the Hankey memorandum known as 'The Boxing Day Paper'.[1] The first step in the production of the

[1] See pp. 80 and 60.

tank was, however, taken by Churchill at the Admiralty. He, as always ready to grasp a new idea, had written early in 1915: 'It is astounding that the Army in the Field and the War Office should have allowed nearly three months of trench warfare to progress without addressing their mind to its special problems.'[2] By March 1916 sufficient tanks had been produced to start forming and training units; Swinton was appointed to command the first unit of six companies, each of twenty-five tanks. Swinton wrote a tactical memorandum[3] which was sent to G.H.Q. and approved by Haig. In it he said:

> Since the chance of success of an attack by tanks lies almost entirely in its novelty, and in the element of surprise . . . these machines should not be used in driblets (for instance, as they may be produced), but the fact of their existence should be kept as secret as possible until the whole are ready to be launched, together with the infantry assault, in one great combined operation.

He thought they should be used on a front of about five miles with 100 yards between each tank. The memorandum stressed the necessity for the co-ordination of all arms in the attack and said:

> The tanks cannot win battles by themselves. They are purely auxiliary to the infantry, and are intended to sweep the obstructions which have hitherto stopped the advance of our infantry beyond the German first line, and cannot with certainty be disposed of by shell fire . . . The principal object of our guns should not be to endeavour to damage the enemy machine-guns, earthworks, and wire, behind the enemy first line, a task they cannot with certainty carry out, and which the tanks are specially designed to perform. It should be to endeavour to help the infantry by helping the tanks, i.e., by concentrating as heavy a counter fire as possible on the enemy's main artillery position and [other known field-guns.]

---

[2] Quoted Swinton, *op. cit.*, p. 103.
[3] Reproduced in full in Edmonds, *op. cit.*, 1916, vol. II, Appx 18.

Cambrai

As has been seen, Haig's desire to renew his attack on the Somme caused him to act against the principles he had agreed and to use tanks when only forty-nine were available and before mechanical troubles had been eradicated. Although, except on this point, there was reasonable accord between Haig and Swinton there are indications that members of the G.H.Q. staff were not so receptive of Swinton's ideas. After the Somme, Elles replaced Swinton in command of the Tank Corps. Fuller, Elles's brilliant senior General Staff Officer, had, long before the advent of the Tank Corps, diagnosed the problem of trench warfare. At the Staff College in 1914 he had preached against the slavish following of any tactical doctrine. He maintained that skill in the use of one's weapons and knowledge of the situation facing one must enable one by common sense to find a solution to a particular problem. Following his own teaching, as a junior staff officer at Headquarters Third Army in 1915, he produced with Captain Townsend a paper known to them as 'the manuscript in red ink'. The conclusion arrived at was '80 per cent of the German Forces on the Western Front are occupying an area some 500 miles long and only 5 miles deep . . . To win the war all that is necessary is to advance 5 miles on a front of say 100 miles. Were this done in the space of a few hours nothing could prevent our winning the war.'[4] There was perhaps nothing unusual in this conclusion, but it was the realization of the time factor and the early recognition of the tank as the means that shows Fuller's genius.

The artillery had been working continually to improve the technique devised for Neuve Chapelle. They received great assistance from the Royal Flying Corps both in photography and in observation of fire. The Royal Engineers also made an important contribution by survey, enabling the exact position of any of our batteries to be accurately plotted on an Artillery Board. Accuracy could, therefore, be guaranteed provided the guns could range, that is by observed fire ascertain the individual performance of the gun and the distance to one of the targets in the area. By these methods the performance of the artillery had greatly improved, parti-

---

[4] Fuller, *Memoirs of an Unconventional Soldier* (Nicholson & Watson, 1931), p. 282.

cularly in counter-battery fire. It was as an answer to the accurate counter-battery fire from the British that the Germans had devised their flexible defence in 1917. But there were two disadvantages inherent in these artillery methods, the business of wire cutting required a long bombardment and the necessity for ranging mitigated against surprise. Four days was considered the minimum time for doing all that the preliminary bombardment had to achieve. In order to prevent the enemy knowing exactly when and where the attack was going to come bombardment was extended beyond that time and over a wide area. The enemy might thus be unaware of the exact time but he was always on the alert. Moreover, the result of a long bombardment, particularly in wet clay soil was absolute destruction of the battlefield, much to the disadvantage of the attacker.

Seeking a way out of these disadvantages Holland had been before his time in suggesting a short bombardment for the Battle of Arras. Although he had convinced Allenby, he had not been able to persuade Birch, and so Haig, that less than four days would suffice. But by November 1917 new techniques had been introduced, particularly sound ranging and flash spotting which enabled special gunner survey units to locate exactly the position of enemy guns and made possible accurate calculation of range and bearing from our own guns. Another essential was a system of calibration which allowed the exact deviation of every gun from normal performance to be recorded. Advances in meteorology enabled correction to be made for wind and barometric pressure. The transfer of the task of destroying the wire obstacle to the tanks was the last link in the chain which set the artillery free from the necessity of opening fire before zero hour. An accurate barrage, neutralization of enemy strong points, and engagement of enemy guns could all be arranged by predicted fire, that is without previous ranging. Thus additional batteries could be secretly moved into an area and an attack prepared without the enemy being any the wiser.

The Battle of Cambrai began where the battle of Arras had left off seven months before. The Germans occupied a position some 6,000 to 7,000 yards deep which included part of the Hindenburg position, prepared at leisure, and behind that a

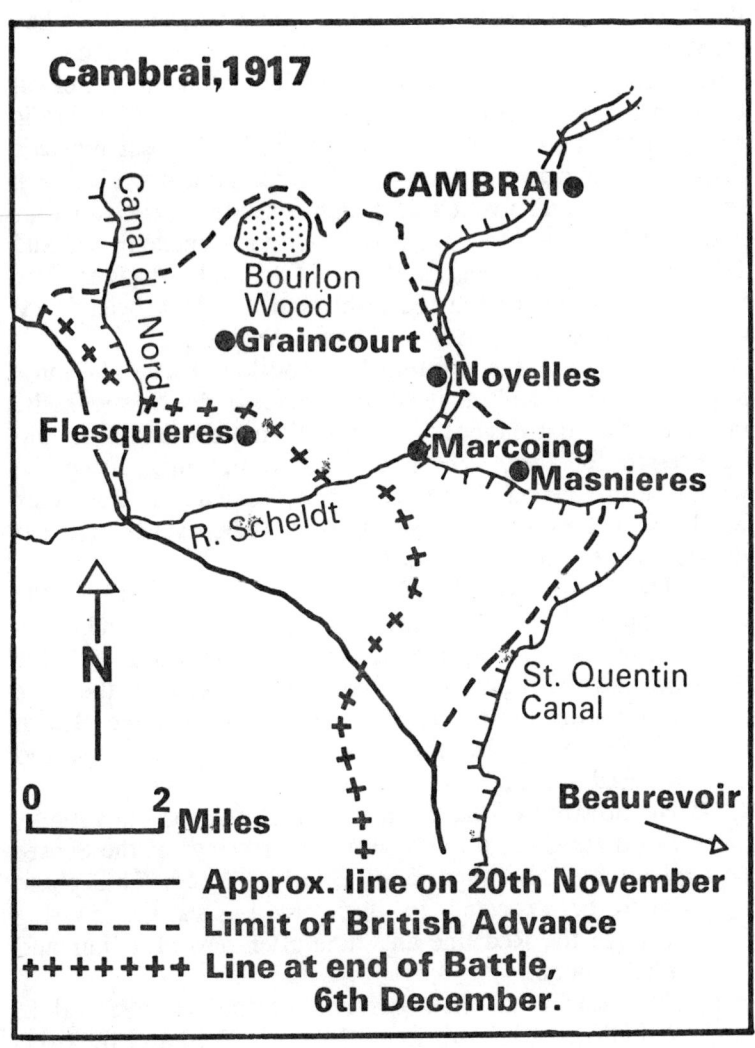

Map 7.

back line which crossed the St Quentin Canal at Noyelles and ran along the ridge to Bourlon Wood. Despite the strength of the German position, this was an area in which an attack had great strategic possibilities. A comparatively short advance would give access to the Douai Plain and a vital area of German communications. The plan finally approved by Haig was that the Third Army would have at its disposal nineteen divisions (organized in four corps), three tank brigades each of 108 tanks, and five Cavalry divisions (the Cavalry Corps and 1st Division). The main attack was to be made by III and IV Corps (six attacking divisions). Two tank brigades were to support III Corps (on the right) and one tank brigade IV Corps, which was also given 1st Cavalry Division.

Haig regarded the capture of Bourlon Ridge as more important than Cambrai itself and directed that it should be taken on the first day. Haig also said that the attack should be pressed for forty-eight hours and if not then successful would be broken off. Byng thereupon decided that the attack would be made on a frontage of about 10,000 yards and that the stages of the attack should be[5]:

(1) The break-through of the Hindenburg Position; the seizure of the canal crossings at Masnières and Marcoing and the capture of the Masnière-Beaurevoir line beyond. This phase involved three objectives, the Blue line taking in the outposts and front line, the Brown line including the support system and Flesquières, and the Red line the Masnières-Beaurevoir defences.

(2) The advance of the cavalry through the gap thus made, to isolate Cambrai and seize the crossings of the Sensée river; and the capture of Bourlon Wood. This phase might be expected to begin by the cavalry passing through the Red line any time after zero plus four and a half hours.

(3) The clearing of Cambrai and of the quadrilateral St Quentin canal-Sensée river-Canal du Nord and the overthrow of the German forces thus cut off.

A large body of heavy and siege artillery was allotted to Third Army, thus allowing a comprehensive and deep counter-battery plan. In all, there were 1,003 artillery pieces to support

[5] Quoted Edmonds, *op. cit.*, 1917, vol. III, p. 18.

the attack. Byng was determined that surprise should be achieved and was adamant that not a single gun should open fire from a new position before zero hour, and that the normal level of artillery fire on the front should not be exceeded before the attack went in. At zero the barrage was to come down on the front edge of the German outpost line and the front trenches of the Hindenburg position, lifts being calculated according to the varying infantry and tank plans of advance. The speed of the tanks was calculated as forty yards per minute on the flat, fifty downhill and thirty uphill. The German trenches in the Hindenburg position were ten feet wide, more than the capacity of the tanks to cross. To overcome this difficulty all tanks carried fascines (enormous bundles of faggots compressed and bound by chains). The artillery detailed to move after the capture of the first objective also carried fascines; batteries were given routes forward to selected positions and arrangements made for hand cutters to widen the gaps in the wire made by the tanks. When the second objective was captured heavy artillery was to be attached in support of each leading division. A Royal Flying Corps Brigade was allotted to support the attack by reconnaissance, bombing communications, and attacks on targets such as batteries, enemy concentrations and strong points, and transport.

Each of the attacking divisions had ten days' training with tanks. The Tank Corps worked out a drill for the attack. The tanks were to lead the infantry through the wire. Each tank section would have allotted to it a definite objective, preferably a rectangle consisting of a forward and support trench and two communication trenches. The leading tank would be directed on to the right of the objective, would penetrate the wire and on reaching the trench would not cross but would turn left down it and engage the occupants. The other two tanks would follow No. 1 through the wire, the second would drop its fascine and cross the trench, turning left down the far side. No. 3 would cross by that fascine, work down the side of the communication trench, drop its fascine and turn left down the far side of the support trench. The infantry would closely follow the tanks, sections in single file. Their task was mopping up, improving gaps through the wire, consolidation, and acting as advanced guard for the further advance. Haig, who visited the training, stressed that the infantry must be prepared to work forward when necessary by use of their own weapons.

All divisions adopted the suggested procedure except 51st (Highland), which devised its own. In this there were to be advanced tank sections which moved four minutes ahead of the infantry to cut the wire. The main body or ' fighting tanks ' were to advance in an arrowhead, the leading tank making straight for the support trench and the outside tanks dealing with the first trench and then supporting the leading tank. Most important change of all, the infantry sections were to follow the fighting tanks not in single file but in extended order. This was bound to cause delays in crossing the narrow lanes in the wire made by the tanks.

The attack went in at dawn on 20 November and in four hours the outpost line and both the strong trench systems that made up the Hindenburg position had been captured. Only one part of the Brown Line was not reached, the village of Flesquières, part of the objective of 51st Division. More than 4,000 German prisoners were taken and the total British casualties did not exceed that number. But much remained to be done if Haig's instruction to capture the Bourlon ridge on the first day was to be obeyed. By dark the attack had lost its impetus; the attack on Bourlon had not been delivered. Marcoing was in British hands but' not Masnières and nowhere had they a footing on the Masnières-Beaurevoir line. Third Army issued orders that both these objectives were to be seized early on the 21st and that as a preliminary Flesquières was to be captured during the night. The object was to get the cavalry through as soon as possible to carry out its original task.

The Germans withdrew from Flesquières during the early hours of the 21st and 51st Division went on to capture Fontaine just under the east of the Bourlon ridge. The 62nd Division on their left also made some progress but by the end of the day IV Corps was still short of the ridge and the brigade in Fontaine was dangerously isolated. On the right III Corps had failed to make any impression on the Masnières-Beaurevoir line. The battle was now almost at the forty-eight hours which Haig had allowed before deciding whether to go on or not. He decided that the attack on the right should be called off but that the capture of Bourlon ridge offered such rewards that efforts to that end should be continued. The attack was persisted in on the succeeding days but though Bourlon Wood, the highest part of the ridge, was captured, the Germans still retained part of the ridge which enabled them to look down on the British artillery positions in the plain. On the afternoon of the 27th Byng called off the attack, a decision accepted by Haig.

On 30 November the Germans launched a major counter stroke to recapture the Hindenburg position. The attack was launched from north and east against the salient made by the British advance. IV Corps from Marcoing to Bourlon Wood held but III Corps and VII Corps on their right suffered a serious reverse. On the first day the British lost some 6,000 prisoners, 103 field and 55 heavier guns. On the morning of 3 December Haig visited Byng and decided that the now deep salient held by IV Corps was too vulnerable to attempt to hold for the winter. He thereupon ordered a withdrawal to the Flesquières line, which was successfully completed by the morning of the 7th. The battle ended with the left of the British position substantially where it had been four hours after the attack was launched, while on their right the Germans had advanced some four miles and had overrun the original British second line.

The failure to capture Bourlon Ridge on the first day was due as much as anything to the check at Flesquières; it is therefore worth retracing our steps to study that action in some detail. Tank experts attribute the failure of 51st Division to capture Flesquières to the refusal of Harper, the commander, to conform to the battle drill suggested by the Tank Corps. Indeed the official history supports this judgment.[6] There are, however, deeper reasons than this, which bring out some of the weaknesses in British tactics once the set-piece phase of a battle was over.

The Germans at Flesquières were in a reverse slope position flanked by guns. The commander of 54th Division, Lieutenant-

---

[6] *Ibid.*, p. 280. A similar opinion was expressed to the author by Liddell Hart in 1965.

General Watter, had made a study of anti-tank tactics and his division had been extremely successful against the French attack at Chemin des Dames earlier in the year, knocking out every tank. Now they proceeded to repeat their success. The guns were neutralized but not destroyed by the British counter-battery fire and when it lifted, the guns were pulled out of their pits and engaged our tanks as they came over the ridge. Twenty-eight of the tanks operating with 51st Division were knocked out by artillery fire. This is possibly the first major indication of the superiority of the well-sited gun over the tank. Disasters of this kind are, however, to be expected in any battle and it is in the subsequent action that the lessons lie. The disaster happened at 10 o'clock and more than an hour later divisional headquarters did not know of it. As a result of a message sent on to IV Corps of 'prisoners coming in from Flesquières' that headquarters reported the capture of Flesquières. Even the headquarters of the brigade concerned did not know until 12.45 p.m. that Flesquières was still in enemy hands. The divisional commander was three miles behind in a position laid down by corps and the brigade commander in his headquarters (laid down by division) where he could not see what was going on. These were days before the introduction of light wireless equipment and the tanks had played havoc with much of the cable, but the incident brings out the tendency to be tied to rear communications rather than the commander being forward where he could influence the battle. This point was well taken by the Germans that winter when they devised their offensive tactics for 1918.[7]

Largely as a result of the ignorance of the commanders as to what was going on, 51st Division did not do all that could have been done to overcome the difficulties. There is no evidence of artillery fire being called for from the heavy batteries specially detailed to provide such support, nor was the reserve brigade of the division used. Co-operation between tanks and infantry broke down after the first check; when the tanks did get into Flesquières they waited in vain for the infantry before withdrawing. When the enemy had again manned their defences the infantry attack went in and was repulsed. It was, however, on the two flanks that the real

[7] *Ibid*, pp. 285, xv.

Cambrai

opportunity of dealing with Flesquières was missed. On the left 62nd Division made good progress and reached Graincourt. They did make some effort to help 51st Division by organizing a tank attack, but their infantry was fully occupied and a combined attack could more easily have been organized by IV Corps with its so far uncommitted reserve, leaving 62nd Division free to press on to Bourlon. On the right flank there was an even better opportunity for 6th Division (in the adjoining III Corps) to help. The opportunity was missed owing to a misunderstanding and a failure of the brigade commander concerned to appreciate the true situation. So far from the divisions on each flank using their success to remove the obstacle to advance, the failure of 51st Division served to restrain their advance.

The 1st Cavalry Division had been allotted to IV Corps; its role, with one company of tanks in its support, was to work on the right flank of the corps after the capture of the Brown Line, and to co-operate on the flank of 62nd Division in their attack on Bourlon. Despite the check at Flesquières there still existed a covered route by which the division could have moved forward. There is no doubt that using their horses to move but fighting on their feet (admittedly reducing their strength by a third) this division might have been able to do much in keeping up the momentum of the attack on the first day.

Although Haig got permission to withhold two of the divisions under orders for Italy,[8] his decision at the end of the second day was not justified by events. After the first day the attack never looked like regaining its momentum. The same attacks made by the same troops, but this time with a depleted number of tanks and with an insufficiently prepared artillery programme, had little chance. The artillery, which by its technical developments had done so much to make the initial surprise possible, showed itself unable to deal with the problems of open warfare. There was now a tendency to rely on map shooting when the situation called for the quick delivery of observed fire. Examples from the action at Flesquières have shown the inability of the infantry and tanks,

---

[8] See p. 74. Four divisions had already left.

and of the command and communications system, to cope with a battle of opportunity.

The lessons in the defensive battle arising out of the German counter-stroke (their first offensive against us since Ypres 1915) are no less important. Indeed this reverse was taken so seriously in high places that the War Cabinet called for a special report from the Commander-in-Chief 'at the earliest possible moment'. They eventually accepted the judgment of Field-Marshal Smuts that 'no one down to and including corps commanders was to blame' but that some brigade and divisional commanders were at fault and that the training of junior officers and N.C.Os required immediate attention.[9] This judgment seems unduly generous to Byng and to Pulteney, the commander of III Corps (he had commanded it continuously since its formation in 1914). Byng had produced a brilliant plan for the offensive but when, partly owing to his own handling of the battle, it had been brought to a halt, he did not seem to have realized how vulnerable to attack his army was. Even the final withdrawal to the Flesquières line came from the intervention of Haig and there is no evidence that before the battle any special precautions were taken to meet the inevitable German attack. Snow, commander of VII Corps, whose 55th Division bore the brunt of the attack, warned his divisions that the attack was about to take place, but neither Byng nor Pulteney did so. The boundary between III and VII Corps was badly chosen, leaving divided responsibility for a likely covered approach. VII Corps was weak in artillery, having no more than eight 6 in. howitzers in addition to the divisional artilleries. In III Corps both the infantry and the artillery were badly sited, particularly in that there was no observation over the canal crossings and that there was much dead ground between the canal and the forward troops. On the other hand IV Corps and 29th Division in III Corps, at the head of the salient, fought staunchly in well organized positions.

Whatever the mistakes of the higher command there is no doubt that the strictures on the state of training of the junior officers and N.C.Os were right. The lack of imagination of the ordinary British soldier is a source of some strength, showing particularly in his willingness to fight doggedly in defence. But

[9] *Ibid.*, p. 296.

it also means that he is apt not to make the necessary preparations to meet the enemy attack and to remain properly alert, unless strictly controlled and ceaselessly guided by his officers and N.C.Os. Haig saw that all was not well in this direction and, after the political ferment was over, he set up a court of enquiry to discover the underlying reason for the poor resistance offered to the German counter-attacks. Maxse was one of the members;[10] in addition to participating in the report he contributed a special note on the necessity for proper battle training of divisions. What he learnt at the enquiry reinforced many of the theories that were at this time evolving in his mind as a result of the 1917 fighting.[11] He considered that:

> The three divisions concerned were weak and distributed on wide frontages, but could have held the enemy long enough to enable reserves to counter-attack . . . Infantry battalions were so surprised as to be ineffectual. The root cause of the trouble was ignorance of the rudiments of successful defence and inexperience in handling sections and platoons as fire units.

He pointed out that:

> It is easier to teach infantry to follow a barrage and take a limited objective than it is to teach them how to make quick decisions in the tactical situations that occur in a defensive battle . . . In other words troops require more training for defence than for offence. Officers require more judgment, they must think and think quickly and make rapid decisions; you cannot do it for them. Therefore practise them now at making rapid decisions and at issuing their orders quickly.

Although Maxse necessarily pointed his remarks to the defensive battle, all that he said reflected also the reason why success in the set-piece battle always came more easily to the British Army than the more fluid conditions in which opportunity for developing the early success existed.

There seems little doubt that much of the trouble stemmed from the idea that we were always on the offensive and that

---

[10] Lieut.-Gen. Hamilton-Gordon was President and the other member was Major-Gen. Pinney.
[11] All quotations are from Maxse Papers.

our forward trenches were merely halting places before the beginning of some new attack. Maxse bears this out in his picture of the battle which he used in his instruction of his corps.

Whenever the Bosche was opposed by rifle fire he made no serious attempt to fight. Infantry posts that held out were unmolested. If the principles of defence in depth had been properly applied, the Bosche would have made little headway after he surprised the front line of resistance. The successful German attack was made on a position we had been holding since March. The Bosche walked through without any preliminary wire cutting and without a barrage. There was no wire sufficient to stop him anywhere, and he attacked with about equal numbers to us, or with only a slight preponderance of numbers.

The British machine-gunners had come in for a good deal of adverse criticism for their part in the defence. Maxse said they had even less idea than the infantry of what to do in the defensive battle. As an example he gave the result of his examination of some of the officers and N.C.Os he had questioned. They unashamedly admitted continuing to fire on their S.O.S lines[12] far behind the enemy attackers when they had within their view, at a range of a few hundred yards, ' Bosche coming over the hill in beautiful order, absolutely on parade, their officers with sticks and maps all slowly and quietly coming over the hill.' After all that the British had suffered from machine-guns in their attacks against the Germans, this seems too much to bear.

Cambrai was not the last time that the British suffered at the hands of a German offensive, but both from the point of view of attack and defence it was a turning point in their handling of the tactical battle. Haig had begun to see that more than weight of material and dogged courage was needed to win battles and from this time began the developments which enabled infantry, tanks, and artillery to take that concerted action which was an essential to victory in battle.

[12] Emergency targets for engagement in darkness, fog, or smoke.

Chapter 7

1918

THE INFANTRY FINDS ITS ROLE

The casualties at Third Ypres and the wilting of the bright flower of victory at Cambrai strengthened Lloyd George's distrust of Haig and Robertson's military judgment, and caused a withholding of the man-power both considered necessary for security on the western front. At the same time the French High Command, supported by the Government, pressed that, the offensive being over, the British Army should take over a substantial part of the French front. The French Army had been nursed back to health by Pétain, but man-power difficulties were causing a reduction in the number of divisions and unless his front could be shortened Pétain would find himself without a reserve. By this time the Supreme War Council at Versailles[1] was beginning to take a hand and on their recommendation Haig agreed to take over a six divisional front opposite St Quentin, stretching to seventeen miles south of the Oise. The British Army had already been reduced from five to four armies when the five divisions and heavy artillery had left for Italy; now man-power shortages caused the reduction of all the British divisions on the western front from twelve to nine battalions, i.e., three battalions per brigade.

The Supreme War Council now attempted to establish a General Reserve in France for the Franco-British and Italian fronts. Neither Haig nor Pétain felt able to provide any formations and they both preferred to make their own arrangements for mutual assistance, unless the reserve could be provided by divisions from elsewhere. Haig expressed the opinion to Robertson that he did not fear that, when the time came to use them, there would be any difficulty in deciding where reserves were most needed. Pétain said that French reserves would be

[1] See p. 74.

at the disposal of Haig, and he expected a similar arrangement to be made by the British.[2]

1918 was obviously a year of crucial importance, both because of the defection of Russia and because it would see the arrival of a substantial United States Army on the western front. The Germans had one last hope of victory with the aid of forces freed from the eastern front, if only they could solve the problem of the offensive battle. The Allies on the other hand inclined to a defensive strategy which would leave the way open for a successful offensive in 1919. The Supreme War Council, in which Foch was undoubtedly the moving spirit, considered that there was no possibility of the Allies gaining a final decision or even a substantial victory in 1918. Pétain agreed with this view but Haig and Robertson both believed that the German all-out offensive would open the way for a counter-stroke which would lead to final victory. There was one point on which Haig and Pétain saw eye to eye and that was the best way in which the Americans could contribute to victory. Both saw attachment of United States battalions or even companies to experienced British and French brigades or battalions as solving their own man-power problems and at the same time giving the Americans essential battle experience. The move across the Atlantic of a number of battalions could be a bonus to, and much accelerate, the programme for the move of completely equipped divisions. Haig had a scheme for incorporating one United States battalion in each infantry brigade, thus restoring his brigades to their original strength. Pétain stressed that unless steps were taken to train divisional commanders and the artillery, the United States could not possibly put an effective army into the field before the next winter; he recommended getting as many infantry battalions into the line as possible so as to provide officers with battle experience for higher command. General Pershing, the American Commander-in-Chief, was unable to agree to any of these suggestions; the Americans desired to see a self-contained army in the field at the earliest possible moment. Pershing did, however, agree that the infantry and divisional troops of his first six divisions should be trained with British divisions behind the line while the Americans trained their own artillery.

[2] Edmonds, *op. cit.*, 1918, vol. I, p. 69.

By March 1918 the Americans had four divisions which had completed their training and were ready to get experience in the line. All were in the French area since this was more easily accessible from Atlantic ports. As the situation became critical, Pershing, delaying his own plans, put all these divisions at Pétain's disposal. Other divisions were training behind the British front; by the beginning of June five were complete and others were in the process of arriving. On the day of the Armistice there were forty-two United States divisions in France.

From the beginning of 1915 the Germans had been mainly on the defensive. They had been working continually on plans for a major offensive on the western front but something in the nature of rescue operations for their allies in the east had always got in the way. Even Verdun and certainly the counter-stroke at Cambrai had been limited in scope. The Germans considered several strategic objectives, but above all they realized that no plan would succeed unless they could achieve a surprise attack on a wide front, supported by large artillery resources and with adequate reserves to exploit success. The final choice of objectives lay between plan ' George ' against the flank and rear of the British in Flanders, and ' Michael ', an attack near St Quentin with its left flank on the Somme and the break used to roll up the British from the south.

Lieutenant-Colonel Wetzell, head of the German operations section, in a full examination of the possibilities came to the conclusion that the difficulties of offensive warfare were such that no single attack would be decisive, so that the best hope for success was an attack at St Quentin to draw British reserves, followed by an attack in the north about a fortnight later directed on Hazebrouck to roll up the British from the north. Ludendorff did not take this most valuable and far-seeing advice. He decided that he had only sufficient forces for a single offensive and that the attack at St Quentin, plan Michael, should take place about 20 March.

Wetzell's paper[3] stressed the necessity for using the winter for thorough training of the divisions earmarked for the attack. He considered that by spring 1918 the German Army ' will be thoroughly fit for an offensive and will be able to show that it is

[3] Translation reproduced in full, *Ibid.*, Appx 20.

superior to all enemies in such an operation'. Discussing his enemies, he wrote of the British: 'We have a strategically clumsy, tactically rigid, but tough enemy in front of us.' He went on: 'The French have shown us what they can do. They are just as skilful in the use of their artillery as of their infantry. Their use of ground in the attack is just as good as in the defence. The French are better in the attack and more skilful in the defence, but are not such good stayers as the British.'

The Germans had learnt much about the tactics of attack from their own successes on the eastern front and from British and French mistakes in 1916 and 1917. The first essential was that the attack should be a surprise and the second that it should be carried out by divisions that had spent the winter specially training for the operation. The impetus of the attack was to come from lightly equipped infantry units thrusting ahead, where necessary by-passing serious resistance and thus disorganizing our defence system. Commanders and rear echelons were to be kept aware of the position of forward troops by a system of light signals. The efforts of all were to be concerned with pressing on where progress was possible rather than with the safety of their own flanks. This system of infiltration was designed to crumble the building rather than batter at strong walls.

Maxse had been able to deduce from his study of the Cambrai counter-attack something of what the German offensive methods would be. In February 1918 when, as part of the extension of the British front, his corps took over the line opposite St Quentin, he taught his divisions something of what they must expect.[4] He wrote:

The Bosche is not going to do like we did last year; we went mad on waves, although we did alter this a bit later in the year. The Bosche is managing his attack another way; he is practising it very much like he did at Cambrai. Creeping Barrage—Smoke Barrage (hand grenades)—then all eggs in the Storm Truppen basket. Storm Truppen are specially trained; they and the machine gunners are the only good people in the German army. Storm Truppen and light machine-guns go right ahead and do not stop for anything. They are followed 200 yards behind by masses of infantry. The Hun

[4] Quotation from Maxse Papers, Feb. 1918.

idea is that these Storm Troops will make holes and continue their advance past our strong points. They have been training to go 12 km (7 miles) the first day. Then the infantry following up behind is to mop up, moving left and right from the breaches thus made, while the Storm Troops move straight on. They expect to paralyse you by the sudden onslaught of their Storm Troops, but you have only to watch properly and have your six men shoot and you will be all right. Thus you will see the hopelessness of trying to defend our line not in depth.

Maxse's answer was defence in depth and prepared and rehearsed counter-attacks. Again, in a lecture in February 1918 he said:

> For two years G.H.Q. has preached depth. The Germans learnt it, the French learnt it. The British agreed to it in principle but not one division in ten carried out depth on the actual ground. But you shall. Many commanders cling to the silly idea that successful defence implies firing every weapon into No Man's Land . . . Each unit in XVIII Corps shall learn to rely on its own resources and not call for reinforcements. If you do call you will not get any. But the higher command will not leave you in the lurch. It alone can decide when and where to strike with a deliberate counter-attack. Meanwhile you must hold on in the forward zone.

When the German attack came on 21 March Maxse's was one of the three corps on which the brunt fell. Maxse's methods in the attack had always been justified by comparative success, but there is little evidence that his defensive teaching bore any fruit on this occasion. The methods he advocated were similar to those which had been successfully used by the Germans, but they depended on two factors which were missing, adequate reserves for counter-attack, and well-trained units. The British Army was stretched owing to the extension of the front allotted to it, and units had not recovered from the exhaustion and casualties of the 1917 battles. In particular the necessity for fighting in sections of six or so men under a junior non-commissioned officer, instead of under close control of platoon or even company officers, proved the weakness at this stage. The Official History has hard words to say on the subject. ' There was a general objection among fighting officers in the distribution of the troops in small packets, the blob system of defence, as it had been called, in derision before the war, for it was not a new theory.'[5] The author goes on to quote an old N.C.O. of 1914 as summing up the new system by saying, ' It don't suit us. The British Army fights in line and won't do any good in these bird cages.'

There were also contrary views and correspondence in Liddell Hart's possession shows Edmonds had an inexplicable bias against Maxse. After every operation Maxse always thoroughly studied what had been done and, in discussion with his divisional commanders and others, tried to deduce lessons for the future. Writing on the March retreat, Williams, commander of 30th Division, said: ' I honestly believe that your XVIII Corps scheme of defence saved the army from a serious disaster. I feel convinced too that, notwithstanding the fog which was an enormous handicap, we could have held our battle zone had our flanks been secure. As far as the Canal du Nord the Hun drove us back; after that we never lost a yard of ground.'[6] Heneker, commander of 8th Division made an interesting observation on the concept of defence ' to the last man and the last round '. In a letter to Maxse he said: ' In order to get the best results one should not hesitate to let all ranks know, if need be, that there are positions to which the defenders may retire when ordered, but that until that order is given, no thought of retreat must be contemplated.'[7]

The truth must surely be that the British defeat in this battle was caused not so much by particular methods of various divisions in the line as by the real skill of the German attacking methods and the failure of the British to work out an effective concept on which the defensive battle should be fought. What Maxse was trying to do in his corps, the German High Command had long since done for their whole army. The German forward zone existed only to delay the enemy; the battle zone was a labyrinth of wire switches intersected with concrete pill

---

[5] Edmonds, *op. cit.*, 1918, vol. I, p. 257.
[6] Maxse Papers.
[7] Maxse Papers.

boxes for machine-guns, proof against all but a direct hit with an 8-inch shell. Above all there were protected positions for adequate counter-attack garrisons. The British had put much more work into the forward zone; there they had a length of trenches which absorbed too many battalions and even then could not be adequately manned. There were no ferro-concrete machine-gun emplacements and no real framework for the defence in depth. The defences on the front recently taken over from the French were in even worse state. From 1915 to 1917 the British had directed all their energies and based all their hopes on a successful offensive. When we knew that the Germans were going over to the offensive the short winter break gave us too little time to perfect a defence system.

The fog on 21 March and each of the succeeding mornings aided the German attack. It exposed the weaknesses of the British hybrid system, since the Germans, whose frontal attacks nearly always failed could always find a way round. By dawn on the 24th the Germans were across the Somme in strength. By the 26th, the day on which Foch was given responsibility for co-ordinating the action of British and French forces on the western front, they had captured Albert. By that day also the German attacks were beginning to lose their impetus and the British Army was beginning to emerge from the ordeal, tired and disorganized, but still ready to fight. The Germans did make further gains towards Amiens up to 4 April, but artillery, ammunition, and supplies were failing to keep up and the German infantry had reached the end of its resources.

As the Amiens attack came to an end Ludendorff turned to the Lys, which Wetzell had wished to include as part of the original offensive. This new attack, launched on 9 April, again in fog, aimed to reach the Lys near Armentières in twenty-four to forty-eight hours. Four divisions attacking the part of the line held by three weak Portuguese brigades went right through to the river on the first day. The 55th (West Lancashire) Division on the southern flank fought a particularly fine action and held. To the north 40th Division taken in flank and rear from the Portuguese position was largely overrun. Cyclists and cavalry, closely followed by 51st (Highland) Division, were rushed up to close the gap. This time the Germans had less resources and had had less time to prepare the attack and the

British positions were held more strongly. By the end of the month the attack was held short of Hazebrouck so again no vital strategic position had been lost. But there had been many anxious moments and on 12 April Haig had issued his 'Backs to the Wall' order. Two days later Foch was officially designated Commander-in-Chief of the Allied Armies in France, although this only confirmed powers he had exercised since 3 April.

Ludendorff determined next to turn against the French so as to draw the Allied reserves away from Flanders and make possible a final blow (Operation Hagen) against the British on the Ypres-Hazebrouck front. The first attack on the Chemin des Dames on 27 May was a considerable success and enabled the Germans next to push out of the salient to the north. The fourth blow came on 15 July on the Marne but three days later Foch was able to launch his counter-stroke with French and American troops, which led, as in the first Battle of the Marne, to the German retirement behind the Aisne. Since 3 June Pétain had begun to see the opportunity that was coming and he must be given credit for the planning and execution of the attack.*

Apart from the fact that it was his counter-stroke that finally destroyed the German ability to attack, Foch truly fulfilled his role as Commander-in-Chief by his handling of the Allied reserves. He had under him the British and the French armies each fighting for their lives, and the Americans eager to join in the battle, but wishing to do so at the earliest possible moment as a national army. Foch moved French divisions, not willingly spared by Pétain, to the succour of the Fifth Army and then, when the weight turned against the French he had British divisions serving under French army commanders and sometimes under French corps commanders. At one moment Foch had taken the last of Haig's available divisions. At the end of the crisis, four British divisions (XXII Corps) took part in the Marne counter-stroke.

The problem of fighting under an Allied commander was not easy because the methods of framing orders was so different. At army and corps level the French were much more detailed and gave a much clearer idea of how the commander expected his subordinates to fight the battle than did the British matter of

* Liddell Hart, *Foch—Man of Orléans* (Faber, 1931), chs XVIII, IX.

fact statement of objectives and frontages. On the other hand at the higher level, orders were more general and left everything to the subordinate fighting the battle. For example, Foch's first action on taking over responsibility was to order Pétain to move reserves north of the Oise, towards the Fifth Army. When Foch visited Gough he simply told him there was to be no further withdrawal and did not mention the movement of reserves.[8] Rawlinson who superseded Gough on 28 March[9] was put directly under French command and did not take easily to French methods. The French habit was to put their own interpretation on orders which showed that their superior was not aware of the true situation. Foch's exhortations and instructions not to give an inch of ground exasperated Rawlinson rather than created a spirit of willing obedience.[10] Later, too, when the Allies turned to the offensive Rawlinson was apt to question Foch's realism even when the orders had been passed on to him by Haig. At unit level, too, the differences were felt. An officer of 15th (Scottish) Division tells how when they were in a French Corps in the Reims sector, the French units on the flank would almost completely evacuate their trenches during heavy shelling.[11] This prudent practice was not allowed under the more rigid British defence system. A battalion commander in the same division[12] gives an even more striking example of the Gallic attitude to orders. The 15th Division received orders from XX French Corps to attack in conjunction with the neighbouring French division. At the appointed time 15th Division found itself attacking alone. A liaison visit to the French division elicited the explanation: 'We will not be ready until to-morrow. We shall be happy if you will join in then.' A successful attack was put in next day by both divisions. When 15th Division left XX Corps, General Fayolle sent Haig a most glowing tribute to its work.[13]

During all these operations Haig remained like a rock; he was poignantly aware of the dangers yet confident that if the Allies acted in concert victory could yet be achieved. Both Haig

---

[8] Edmonds, *op. cit.*, 1918, vol. I, p. 543.
[9] This was a political decision and subject to a strong protest by Haig.
[10] *Ibid.*, vol. II, p. 487.
[11] Lt.-Col. Stuart Cameron (then a subaltern in 6th Camerons) to the author.
[12] Brig. A. C. L. Stanley Clarke then commanding 10th Scottish Rifles, to the author, Sept. 6 1967.
[13] Reproduced in Edmonds, *op. cit.*, 1918, vol. III, Appx XVI.

and Pétain saw that if the Germans could separate the French and British Armies the war would be lost. Haig believed that the French could prevent this by the use of their reserves to hold to the British flank in front of Amiens. Although the French were not at this stage being seriously attacked, Pétain believed that the danger to Paris was such that Haig must be prepared to draw close to the French even at the risk of losing the Channel ports. Possibly Pétain misunderstood Haig's intentions and believed that he would withdraw not westward but northward; certain it is that on 24 March he admitted to Haig that he contemplated abandoning touch with the British right flank.[14] It was this understanding of the fact that Pétain contemplated defeat, confirmed at the Doullens Conference on 26 March,[15] that caused Haig to demand a Generalissimo and to declare his willingness to serve under Foch. He suggested, and Pétain agreed, that Foch should have full authority over all the operations on the western front.

The course of the defensive battles, both before and after Foch assumed command, shows clearly that a commander without a reserve has no possibility of influencing the battle. On 21 March, owing to the extension of the British front, Haig had only eight divisions in G.H.Q. reserve for his 126 mile front. He placed two behind each of his four armies (there was no Fourth Army). The Fifth Army had only one division and one cavalry division in its own reserve, the Second, two divisions and each of the others one only. As his reserves were committed to the battle Haig took steps to create new reserves but he could only do so by calling on the very part of his front that was to be the subject of the next attack. Thus, as the Amiens threat subsided and the Germans turned to the Lys, the situation was even worse because of the number of divisions no longer fit for battle without rest and reinforcements. Before the German attack came Haig knew that 2nd Portuguese Division was weak and strung out, but he could not relieve it except by using his one completely effective division still in reserve. Apart from that all he had was nine divisions which had been badly mauled in the Amiens battle and were in the

[14] *Ibid.*, vol. I, p. 450.
[15] *Ibid.*, vol. I, p. 539.

process of being made up with reinforcements, some of them still on the way.

As has been shown Foch was willing to use his authority with Pétain to move French divisions to the Amiens front. They were slow to move and in the first place infantry arrived without artillery, but the link between the two armies was saved. When the German attack switched to the Lys, Foch was more reluctant to move because he did not altogether discount Pétain's fears that the Germans were trying to draw reserves away from Paris. At first, therefore, Foch was determined not to place French troops further north than was necessary to intervene in a battle for Arras. There was a clear tussle of wills; Haig insistent that the destruction of the British Army was the immediate object of the enemy, Foch soon accepting this view but carefully measuring the extent to which the British Army could work out its own salvation, and seeking always to keep in hand the reserves with which he could eventually gain the initiative. Nevertheless, before April was out Foch had gone some way to meet Haig's demands and the French had five divisions and three cavalry divisions in Flanders.

When the Germans switched the offensive to the French front Haig and Pétain changed roles in the pull for the Allied reserves. Haig believed, rightly as subsequent knowledge shows, that the Germans were trying only to disperse the Allied effort before seeking a final decision in Flanders. But the German successes created a dangerous situation on the Aisne and Foch rightly called on Haig for the return of the French divisions, for the move south of American divisions, and for British divisions as well. Haig protested vigorously but obeyed. Haig was in a difficult position, as indeed Pétain had been. Although Foch was in command, Haig was still charged with the safety of the British Army and it was his duty to appeal to the British Government if he received orders which seemed to imperil that safety. Once it was accepted that Foch could order the move of British forces away from Haig's command the responsibility obviously lay with Foch. On 21 June Haig's original instructions were slightly modified; he was ordered loyally to carry out the instructions of the Allied Commander-in-Chief and, although the right to appeal to the British

Government was retained, the hope was expressed that the necessity would never arise.[16]

The measure of Haig's performance during these difficult days is that he not only succeeded in retaining the confidence of his own army but that he grew in Foch's estimation. The relationship between Foch and the French G.Q.G., particularly with Pétain, was certainly less easy than that between Foch and G.H.Q. This confidence and mutual understanding was to stand the Allies in good stead when they turned to the offensive.

While the Battle of the Marne was still in progress, Foch's mind was turning to a British offensive; his first thought was for an attack in the Festubert-La Bassée area. Haig saw no profit in being again engaged in the water-logged area of the Upper Scheldt and persuaded Foch that an operation in the Amiens area would be much more profitable. A preliminary operation at Hamel (to which further reference will be made) had shown that the German defences were poor while the firm open country, free from shell cratering, was favourable for the action of tanks and cavalry. Haig already had Rawlinson, whose Army was again numbered Fourth, working on plans and Foch not only accepted the idea but put the French First Army under Haig to co-operate on the southern flank.

One of the lessons of the defensive battles, particularly the March retreat, was that the British Army was not adequately trained for open warfare. The type of operation that the B.E.F. had carried out so well in August 1914 was beyond the ken of the greatly expanded army. This had been seen when the army had tried to follow up the Germans as they retreated to the Hindenburg Line in 1917, and again at Cambrai, but nothing had been done about it. Now there is evidence that senior officers were really giving it attention. A paper by Uniacke, Major-General Royal Artillery in Fifth Army, brings out the faults and looks for remedies.[17] Uniacke had a refreshingly broad outlook and saw the faults on the command side as well as in the junior leadership. The paper dealt with a number of technical points in the selection of a defensive position and the conduct of a withdrawal, in which too little attention had been given to artillery observation and fields of

[16] Secretary of State's Instruction reproduced in full. *Ibid.*, vol. III, Appx IX.
[17] Great War Papers in R.A. Institution.

fire for machine-guns. He pointed out the great superiority of the Germans in the use of machine-guns. Allied counter-attacks were almost always stopped by hostile machine-guns and the chief failure of the British throughout the war had been in the tactical handling of machine-guns. We had also failed to produce a good mobile trench artillery; the German use of mortars was far superior to ours.

This time Haig acted. He decided to set up a special branch at G.H.Q. to study the problem of training for mobile war and to provide a team which could visit each division in turn to organize the instruction. He appointed Maxse to become Inspector General of Training at its head. Uniacke became the Major-General Royal Artillery, Hollond, who had been Maxse's senior General Staff officer in XVIII Corps, was chief of staff and co-ordinator. There were three other Brigadier-Generals as 'Inspectors', Dugan, Marshall and Guggisberg, all of whom had seen hard fighting in command of brigades. One of the junior staff officers was Barrington Ward, later editor of *The Times*; he was used to edit the pamphlets and instructions. The Inspectors used to visit the battlefields with their staff officers to make their observations and recommendations on training. This had the advantage of having someone free from command responsibility or staff work actually studying what was going on. Evetts, who was staff officer to Dugan, writes:[18]

> I have a vivid recollection of watching with Dugan the attack of 9th Division (in which I had been a Brigade Major) on 28 September 1918, which went like clockwork. I also can see, in my mind's eye, the 'blobs' and snake like infantry formations in the attack . . . the ground was quite unsuitable for 'waves'. There was only one gun to about 45 yards and 9th Division had to rely chiefly on smoke for the barrage, and so the guns fired two rounds of smoke to every round of H.E. Mules with ammunition etc were used very close behind the infantry and due to the way the Sappers and Pioneers repaired the roads immediately behind the infantry the Divisional Artillery was able to move forward at 0800, zero having been at 0525.

[18] Letter to the author, Aug. 1967.

The Battle of Hamel is important in that it shows Monash, commander of the Australian Corps, as an outstanding organizer of the offensive battle. The methods employed bring to its final stages the technique used at Cambrai and served as a model for the major offensive on 8 August. Monash had been a Civil Engineer and become a Colonel in the Australian Citizen Forces in 1913. His rise to high command came from hard experience. He had commanded 4th Australian Brigade during the Gallipoli Campaign and had there shown his organizing ability but had not been altogether successful in coping with the exigencies of battle. Contemporary criticism rates Monash's brigade the least effective of the Australian contingent, and an account of the campaign says: ' Monash had prepared plans with meticulous attention to detail, but did not possess resources to deal with situations when they went wrong.'[19] Nevertheless those who know the Australians and their robust attitude to their commanders will temper this judgment by the fact that Monash was later selected to command the newly-formed 3rd Australian Division. Under him this division fought at Messines Ridge and took an outstanding part in the Battle of Broodseinde, towards the end of the Passchendaele battles. When Birdwood left the Australian Corps to command Fifth Army Monash succeeded him. Possibly Monash was one of those men who gets better the higher he goes, but he was certainly not one to climb on other men's abilities; he understood the ingredients of battle right down to the part to be played by the individual infantry soldier.

Monash believed in tackling a problem from first principles. In his own words:

> It did not take me long to learn that the only way to carry out the responsibilities of command were first to inject optimism into a creed for myself, and for all my brigades, arms, and departments, and secondly to try to deal with every task and every situation on the basis of simple business propositions, differing in no way from the problems of civil life, except that they are governed by special technique. The

---

[19] Rhodes James, *Gallipoli* (Batsford, 1966), pp. 271-2.

main thing is always to have a plan, if it is not the best plan, it is, at least better than no plan at all.[20]

The Battle of Hamel was a limited attack on a front of about four miles to a depth of 2,500 yards. Ten battalions of infantry and sixty tanks took part. This was the first battle in which American troops took part with British forces and Rawlinson chose the appropriate date of 4 July. In allotting a United States battalion to the Australian Corps, Haig fulfilled Pershing's wishes in ordering that it should be used intact to fight as such. Monash protested that inexperienced troops should not participate unless they could be incorporated into Australian units. Despite the weight of opinion from above, Monash was adamant, visiting Haig to get his way. The four American companies were used by platoons in the several Australian battalions. The idea that tanks should be used in this attack came from Rawlinson; he wrote in his diary:

> I introduced several novelties into the attack. The Australians had no confidence in the tanks after Bullecourt (Apr. 1917), where the tanks went wrong and did not turn up in time, and ended by shooting at them; so, when I proposed using tanks this time, the suggestion was not welcomed. However, I made them do it. We practised with the tanks daily, and showed the Anzacs what they could do, and how to use them.[21]

The tanks used were the new Mark V much more reliable than the previous models.

Monash ensured that all the troops taking part should train specially for the operation, and the combined training of infantry and tanks was of great value. There was no preliminary bombardment and every possible step was taken to ensure surprise. The infantry formed up on tapes and tanks moved forward to them under cover of the normal harassing fire programme. The attack went in when it was just light enough to distinguish friend from foe at twenty yards. The tanks were put under the infantry battalion commanders and preceded the

[20] Quoted in an article on Monash by John Terraine in *History Today*, 1966.
[21] Maurice, *Life of Rawlinson* (Cassell, 1928).

infantry under cover of a heavy creeping barrage. The attack was a complete success; the tanks played an important part but the feature of the attack was the skill with which the infantry took advantage of them and, when necessary, fought their way forward with their own weapons. An account of the battle was produced for issue down to all battalions and batteries as one of a series of training pamphlets.[22]

Although Hamel was a complete success it was only a limited attack, and the problem to which Monash now turned his mind was that of maintaining the impetus of the attack. As divisional commander at Passchendaele he had been faced with the same difficulty. He had experimented with a system of leapfrogging battalions so that fresh troops were available for subsequent objectives. Now he determined on an extension of this idea.[23] In preparing, under Rawlinson, for the Amiens offensive he saw that the problem was to get right through to the enemy artillery on the first day. He decided to do this in three phases. A set piece attack would be necessary to break into a depth of 3,000 yards; provided fresh divisions could be at hand, this would give the opportunity for open warfare operations to a further 4,500 yards, and these divisions could use their own reserves for exploitation to the gun lines, a further 1,500 yards. The difficulty of this type of pattern had always been the time factor. The passing of one division through another had been such a ponderous manoeuvre that the enemy had time to close the gap or to counter-attack. Moreover the move up of a division from the rear meant a long march which brought tired, not fresh troops to the battle. Now Monash devised the idea of holding the front lightly, with a brigade of each of the assaulting divisions on a rather extended front. Behind them would be, not the other two brigades of their own divisions, but the divisions destined for the second phase or open warfare. Behind these were the two brigades per division for the first phase assault. Thus the troops for the critical mobile operations were already at hand, their commanders well forward where they could see the way through. The concentration was to be undertaken in absolute secrecy and surprise ensured by arranging artillery and tank

[22] Notes on Recent Fighting, No 19. Issued by the General Staff, August 1918.
[23] Monash, *The Australian Victories in France* (Australia, 1936), pp. 61 *et seq.*

support on the lines developed at Cambrai and Hamel. An intensive programme of combined training was arranged; infantry and tanks who were to fight together trained and worked out their problems in common. Not only were artillery batteries, of whose presence the enemy was ignorant, available to open fire accurately at zero hour, but a comprehensive programme of their move forward for the mobile phase had been worked out and practised; thus all the problems of timing the move up of the horse teams and of ammunition supply had been tackled in advance.

The Australian Corps was only one of the three in Rawlinson's Army. Haig's plan was that Fourth Army and the French First Army should capture the German line, known as the Amiens Outer Defence line, between the rivers Somme and Aire, an advance of some six miles on a front of about ten miles. They should then be prepared to exploit towards Ham, and for this purpose the Cavalry Corps and some battalions of whippet tanks were put under Rawlinson's command.

The attack went in on 8 August, which was, like 21 March and 9 April, a morning of mist. The mist caused some confusion among both attackers and defenders, but certainly assisted well-trained and well-led infantry in the attack. The Australian Corps captured all its objectives soon after mid-day, all had been achieved in half a day's fighting. The Canadian Corps, on their right, was almost equally successful, but on the left III Corps succeeded in capturing only the first of their three objectives. The French First Army was slower than the Canadian Corps and caused some difficulties to their right division, but by the end of the day they had made substantial progress, although they had not captured their final objective.

8 August was indeed the ' Black Day of the German Army ' but it is questionable whether full advantage was taken of the Australian and Canadian success. Haig was critical of Rawlinson's lack of boldness. He handled the cavalry with caution and did little more than push out a light screen ahead of the infantry. The Australian success might have been used to bypass the opposition holding up III Corps. Rawlinson did not commit the one division he had in army reserve to exploit either the Australian or Canadian success. The 17th Armoured Car Battalion, working under the

Australian corps, showed by a strike deep into the enemy lines what might have been achieved by mobile troops boldly handled. There was a good deal of disorganization in rear of the enemy and the air reported all roads leading to the Somme crossings as being crowded with transport. Our air forces, in conditions of low cloud which gave some assistance to enemy fighters, were used almost exclusively to try to destroy the bridges over the Somme, a task in which they nowhere succeeded. There seems little doubt that they would have been more effective and have suffered fewer casualties if they had been used against enemy columns.

The Battle of Amiens was one of a series of thrusts by which Haig destroyed the capacity of the German Army to fight. On 22 August the offensive was extended to the north by Third Army and later by First Army. At the same time, as part of Foch's plan, the French Tenth and Third Armies were making good progress in the Soissons area. Thus the Allies were attacking all along the front from Soissons to Arras. On 22 August Haig sent a telegram to all his Army Commanders and the Cavalry Corps Commander. He pointed out that the methods of attack with limited objectives no longer met the conditions. The German Army no longer had the means for counter-attack or for holding a continuous position. It was not now necessary to advance in regular lines step by step but divisions should be given distant objectives which must be captured even if flanks were left exposed. Vigorous effort where the enemy was weak would cause the strong points to fall, and would cost less than more deliberate methods.[24]

This phase of the offensive saw another striking success by the Australian Corps in the capture of Péronne. The Germans were in a strong position behind the Somme, which west of Péronne makes a right-angle turn to the south. The Fourth Army had not followed up quickly enough to seize the bridges before they were destroyed. The river here is virtually a marsh 1,000 yards wide intersected by rivulets, and the bend in the river is protected by Mont St Quentin commanding both reaches of the river. Monash used his position north of the river in order to turn the enemy defences. To achieve this, Mont St Quentin was captured by a brilliant feat of arms against

[24] Reproduced in full, Edmonds, *op. cit.*, 1918, vol. IV, Appx XX.

superior numbers. There was no time to prepare an artillery barrage so, after every available mortar and howitzer had fired on the hill for thirty minutes, the infantry by sheer skill working through old trenches and from shell hole to shell hole, captured the hill.

While the German flank on the Somme was thus turned, the Third Army turned the other flank at Arras, with the result that the enemy withdrew from the Lys salient and was again back on the Hindenburg Line. The greatest blow of all came at the end of September and the beginning of October, with the storming of the Canal du Nord, the shattering of the Hindenburg Line, and the Second Army success at Ypres which endangered the German hold on the Belgian coast. Continued operations at Courtrai and Maubeuge completed Haig's strategic aim by driving a wedge into the German Army, pushing the two parts each side of the Ardennes.

During all these operations Haig called the tune, and it was his direction of the British Armies that enabled Foch to gather the fruits of his strategic vision. Foch was not so successful in moving Pétain to take full advantage of the gradual deterioration in the German will to victory. Indeed it was east of Reims that the strategic opportunity for rolling up the German line existed, but the French had borne too much of the weight of the war in the first two years, and the Americans had not yet the experience to exert their full weight. Two American divisions played a part in the operations of Second Army from September onwards, but the baptism of fire of the United States Army as such was the special operation to cut out the St Mihiel salient, which began on 12 September. The salient, a legacy from the original German attack in 1914, bit sixteen miles deep into the French position south of Verdun. The operation was well within the capacity of the American First Army, in fact Pershing wanted to go much further and exploit towards Metz. However, Foch was persuaded by Haig that the French and American effort should be concentric to the British attacks. The St Mihiel attack was therefore treated as a limited attack and the Americans then directed to make hurried preparations for an attack, in conjunction with the French Fourth Army, in the direction of Argonne-Mézières. This attack went in just as Haig was developing his assault on the Hindenburg Line and made disappointingly slow progress. It was not until the beginning of November, by which time the German Army was already beaten, that the Americans really got into their stride. At the last moments of the war there is thus another illustration of the difficulties which are involved in the rapid expansion of a small regular army into a nation in arms.

Chapter 8

HAIG AND HAMILTON

A STUDY IN COMMAND

The British military commanders in the Great War, particularly Haig, bore greater burdens than ever before in history. Marlborough was the genius of the alliance against Louis XIV and Wellington played an important part in bringing about the downfall of Napoleon, but in each case the British were taking a subsidiary part in the land campaign against a dictator who had already established himself in the hegemony of Europe. Defeat at Oudenarde or at Waterloo, serious as either would have been, would have re-established a situation which had already existed and would have left British power relatively unimpaired outside the heart of Europe, and still left the possibility of a later resurgence. A German victory in 1914-18 would have meant the end of the British Empire, not an end such as has since occurred by transition and consent, but an end by conquest which would have removed us from all influence in the affairs of Europe. Yet, when victory came, little credit was given to British generalship; if any general had been given a place among the Great Captains it would have been Foch, or even the defeated Ludendorff.

Douglas Haig and Ian Hamilton were the two British commanders in whose hands it lay to shape the outcome of a vital campaign. Allenby designed and executed a campaign with a brilliance which eluded the other two, but he must receive less consideration because of the smaller part his campaign took in the war as a whole. Moreover, his problem was easier in that he had room to manoeuvre, an advantage which was denied to the other two.

Haig was never completely independent in command, but, from mid-1916, when the British Army had grown to its full strength and the French were feeling the weight they had borne for two years, he was master of the fate of the B.E.F. Even in the unfortunate interlude under Nivelle, Haig carried out the operation at Arras that he wanted. When fully subordinated to Foch, he exerted perhaps a greater influence on the campaign as a whole than at any other time. Haig never had the confidence of Lloyd George; he suffered less from this than might have been. He did have the confidence of most others in the government, and of the nation, and above all of the army. Moreover, he was insulated from the politicians by Robertson who felt the edge of the differences between soldiers and statesmen. When Robertson had gone it might have been very different, for indeed Haig had a profound distrust of Wilson, but the German offensive came so soon, and Haig was so obviously the man to deal with the crisis, that he remained firmly in command.

Hamilton was militarily independent in command. The C.I.G.S. was at that time a cipher and Hamilton dealt direct with his old chief, Kitchener. Hamilton was thus much more closely affected by the political field, and also by naval considerations. It was not Hamilton's fault that, even before his force was put into the field, strategic surprise had been lost. His former relationship to Kitchener and his great loyalty to him were something of a disadvantage. He accepted without protest the limiting of his force to 75,000, although he knew this gave him an inferiority in numbers to the Turks and left him without the strategic reserve necessary for his hazardous operation. His loyalty in refusing to add to Kitchener's political difficulties by asking for more can be admired, but as a commander he cannot be excused for failure to insist on having the necessary reserves, particularly as they could have been made available from Egypt.

The conclusion of this survey of the strategical control within which both commanders had to work must be that, although they suffered the fetters and the frustrations which are inherent in modern high command, neither was unduly hampered by political or War Office control and that, if the tasks set their armies were attainable, the result lay in their own hands.

The two Scotsmen were certainly different in character and military outlook. Hamilton, the older by eighteen years, was an infantryman with all the brilliance and dash usually associated with the cavalry leader. He had indeed made a great name in South Africa with the mounted infantry. Haig was dour and determined and, although a cavalryman, had all the attributes of a dogged Lowland infantryman. Imagination and deception were the basis of Hamilton's military skill. Haig would see what had to be done and set out deliberately to do it. Hamilton believed that no plan could come to fruition unless subordinates given a task were left to their own initiative to carry it out. Haig used his implacable will to impose the desired course of action on his subordinates. Both saw the over-riding importance of morale, the soul and well-being of an army expressed in the will to win. Hamilton sought to destroy the will of the enemy by surprising and outwitting him; Haig believed that the war could only be won by the direct defeat of the German Army. Both were optimistic in outlook and believed that a plan they had decided on must succeed, but it worked in a different way. Hamilton was always questioning while Haig, open-minded in deciding on his plan, once he had made up his mind went far beyond proper resolution and optimism in holding to the set course. The differences can be seen in all that the two commanders did and achieved. Where the one failed, or partly failed, it seems that what was lacking was just that quality which shone so brightly in the other.

Hamilton's characteristics can be traced in his conception and plan for the capture of the Dardanelles. The way in which strategic surprise had been lost has already been related. But, despite the two months warning given to the Turks by the naval bombardment, Hamilton did not despair of achieving tactical surprise by ruses and deception. An attack on the isthmus of Bulair was out of the question since the Turks had made formidable preparations to meet a threat in that vital area. Hamilton decided on a double thrust to seize the peninsula further south. The Anzac Corps, commanded by Birdwood was to land north of Gaba Tepe and strike across to the Narrows, while 29th Division commanded by Hunter-Weston was to land at Cape Helles and seize the commanding height of Achi Baba. The difficulty of the capture of Achi Baba was not overlooked, although the fighting value of the Turk in defence was undoubtedly under-estimated. There was less than one Turkish division south of Gaba Tepe and the success of the Anzac landing would cut off Cape Helles from Turkish reserves.

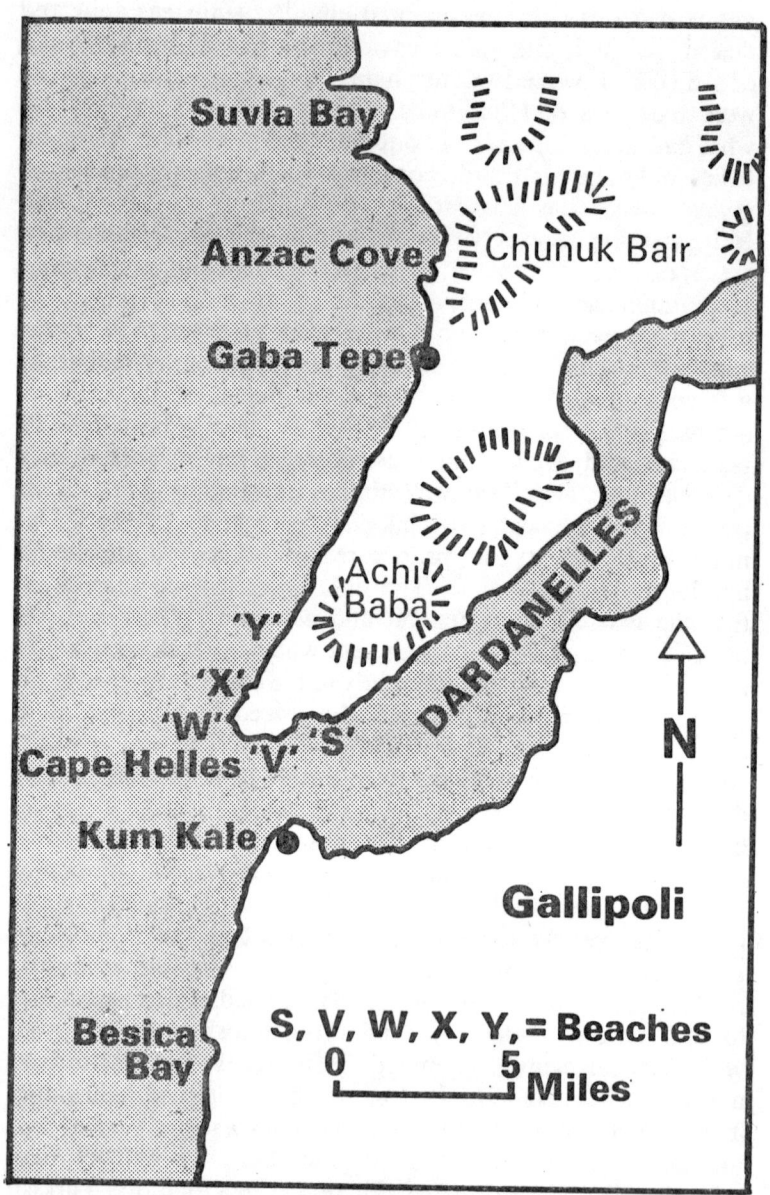

Map 8.

Moreover, the plan allowed for deceptive moves towards Bulair and Besica Bay as well as a diversionary landing by a French force at Kum Kale, which would engage the garrison on the Asiatic side. The landings near Cape Helles were well placed to outflank the Turkish defenders; in addition to the three landings at the tip (V, W, and X Beaches) there were landings in the east (S Beach) and in the west (Y Beach). Every known artifice was used to facilitate quick landing of the troops necessary to cover the disembarkation of the main body, notably the use of the steamer ' River Clyde ' to run ashore and land 2,000 men while the enemy were still engaged by the first parties landed in the naval tows.

The brilliance of the plan may be judged by the fact that at only two of the beaches, V and W, was any serious opposition offered to the landing. At Anzac Cove, unhappily, the difficulties of navigation in pitch darkness had led to the landing being made a mile too far to the north, and had brought the Anzac Corps ashore in an area of precipitous ridges and tangled ravines. But there and at S, X and Y beaches the first and most hazardous stage of the operation was accomplished almost without loss. Yet, despite the success, neither the Anzac Corps nor the 29th Division came within measurable distance of reaching the objectives laid down. Indeed the first day's objective for the 29th Division, Achi Baba, was never attained throughout the campaign.

At Anzac the failure to take advantage of a fleeting opportunity was due partly to the difficulty of the ground but also to some extent to inexperience and to indifferent leadership, a fault which was seldom again to appear in the Anzac Corps. At Helles it was due to the fault, so often shown in France, of hammering at the barred door when an easier way in existed. Hunter-Weston was intent on making good the landings at V and W Beaches, when he could have achieved all that he wanted by using his footing at S and Y Beaches. Hamilton saw the opportunities that offered, especially at Y Beach. While it was still possible to do something he sent a message to Hunter-Weston that trawlers were available to get more men ashore at Y Beach. But Hunter-Weston was intent on the grim battle being so gallantly fought at W Beach (Lancashire Landing). The opportunity at Y Beach did not recur, the commander there, unconscious of the role his landing could play, withdrew his force on the morning of the 26th. During his twenty-nine hours ashore he had not received a single message from 29th Division.

The obvious reason why Hamilton did not exert more influence on the battle once the landings had gone in, is that he had no appreciable reserves with which to act. The extent to which this was his own fault has already been indicated. But he did have two French brigades and he offered them to Hunter-Weston to put ashore at Y Beach. The only answer he got was a request for one of them for W Beach. Hamilton's failure to insist on what he saw was the right course exemplifies his refusal to interfere with the man on the spot. Hunter-Weston was a proved fighting soldier, trusted and loved by the commanders and men with whom he served. Hamilton felt he must be allowed to get on with the task that had been entrusted to him. Hamilton's views on the relationship of a commander to his subordinates is best described in his own words:

> The true commander, then, knows the truth of that great maxim: if the subordinate never makes a mistake, he never makes anything . . . Against the danger of a subaltern's mistake in the execution of his own job, it is but fair to set the risk of a meddlesome superior failing himself in the performance of another's business . . . The superior is too busy doing someone else's job to attend to his own—too busy with the parts to give his mind to the whole. And, of course, what is true of the subaltern and his superiors is equally true at all levels throughout any Service.[1]

The very nature of the plan required all the available forces for the several landings, and orders had gone into great detail as to what was to be done at each. Hunter-Weston's orders for the landings were admirable but he left too little to the initiative of his subordinates for the subsequent action. For example the commanders on the two flanks, at S and Y, were told that after the capture of their beaches they were to wait for the attack

---

[1] Hamilton, ed. Farrar-Hockley, *The Commander* (Hollis and Carter, 1957), p. 53.

from the south before joining in the attack on Achi Baba. This rigidity was inexcusable in the light of the possibility that some but not all of the landings would be successful. It was all the more disadvantageous in the light of the communications difficulties. With five landings Hunter-Weston would have been in no position to command in the early stages from any one of them. As it was, the ship from which he commanded was never more than a mile from his front line but, despite the efforts of the Royal Navy, communications between ship and shore were not effective. The location of Hamilton's headquarters in H.M.S. *Queen Elizabeth* was not without serious disadvantages. Although in theory such a headquarters gave him the opportunity of moving more quickly and more closely to the widespread areas of the battle than would be possible on land, the requirement for the guns of the *Queen Elizabeth* to support troops on land tied him for hours to the Helles area.

That Hamilton had the courage of decision as well as military insight is shown by events at Anzac. Here Birdwood, faced with a rapidly deteriorating situation and the conviction of his two divisional commanders that evacuation was inevitable, put the facts to Hamilton and asked for his decision. Without hesitation, Hamilton came to the right decision that the Anzac Corps must ' make a supreme effort to hold their ground '. They did. There is no doubt that the failure to achieve more at Helles was due, not to the very real difficulties which have already been enumerated, but to Hunter-Weston's blind concentration on the southern beaches, and to Hamilton's obsession with the freedom of his subordinates to run their own show.

Hamilton's failure to impose on his subordinates his own tactical conception and to galvanize them into the action required comes out even more clearly at the Suvla landings. Again there was a brilliant plan and a brilliant surprise was achieved. The basis of the plan was a new attack on Chunuk Bair and the turning of the flank of the enemy to capture a line facing south across the peninsula from Gaba Tepe. In order to overcome the difficulties presented by the restricted area in which the Anzac Corps were working, and to widen the scope of the operations a corps of two New Army divisions was to land at Suvla Bay and seize the line of hills on the northern flank of the Anzac Corps. The early attacks by the Anzac Corps were entirely successful and the landing places at Suvla were gained without loss. A magnificent exercise in fighting command by Mustapha Kemal prevented the Anzacs from reaping the full advantage of their early success, but the fruits of the landing at Suvla were thrown away by the complete failure of the corps and divisional commanders, and many of the subordinate commanders, to grasp the opportunity. The New Army divisions were magnificent material but they were inexperienced. They were in the main commanded by officers who had already retired before the war and who now found themselves in higher positions than they would otherwise have reached. Younger men might have risen to the occasion, but these were hardly the men to lead the fine officers and men of ' The First Hundred Thousand ' whose first experience of battle was to be a hurried dash across country by night to seize an unreconnoitred line of hills. In Hamilton's defence it must be said that he had asked for young and experienced corps and divisional commanders, but Kitchener had not listened. So it was that experience at second hand induced a spirit of caution by false lessons from the tale of slow and deliberate advances on the western front.

Hamilton is to blame on two counts. In the preliminary discussion of the plan it was clear that Stopford, the corps commander, was too cautious in his approach. Hamilton allowed some watering down of his intentions previously clearly expressed in an instruction to capture a certain essential group of hills before dawn.[2] Then, after the landing Hamilton was early on the scene and visited the corps and both divisional commanders, and was well forward under fire to see what was going on. Whatever his feelings at the inactive spirit he found, he appears to have said little. It was not until after visiting Anzac Corps and returning to his own headquarters that evening that he expressed his doubts in a letter to Stopford and tried to impose his will.[3] There seems little question that Hamilton had had it in his power that morning to avoid

---

[2] Aspinall-Oglander, *op. cit.*, vol. II, pp. 149-151 deals with this in detail and reproduces the first and the final instructions issued by Hamilton in full at Apps. 2, 3.
[3] *Ibid.*, p. 298. Letter in full.

Haig and Hamilton 155

failure by insisting on his subordinates doing what he saw clearly must and could be done.

It seems clear that Hamilton's imaginative approach to the problems of battle, his resilience of mind and of spirit, was marred by the absence of that ruthlessness towards his subordinates which is one of the essentials of a great commander. A recent study of the campaign[4] has thrown doubts even on the first quality by questioning the part which Hamilton played in the conception of the two great plans. The author attributes the plan for the original landing to Lieutenant-Colonel Aspinall, head of Operations Section at G.H.Q.[5] and to Guy Dawnay, a member of his staff. The plan for the Suvla operation is chiefly attributed to Birdwood. This is unjust to Hamilton and seems to overlook the process by which a plan is made and the part played by the commander, his staff, and his subordinates. The fact that a commander accepts an idea or a series of ideas from one of his staff is to the credit of both. The putting forward of the idea, even the essential brilliant germ, is only a small part of the plan. The responsibility is still with the commander, and all the stages of working it out and giving it final life depend upon him. It is true, too, that Birdwood first suggested the stroke from the Anzac positions out of which blossomed the Suvla plan, but all the later work, including the addition of the Suvla landing went on under Hamilton's vital direction; here, too, Aspinall and Dawnay played a most valuable part.

Hamilton wrote much on command and leadership. His own list of the requirements of the commander is long. He thought the 'supreme' commander must be a master in strategy and in tactics; prepared to take a gamble when the prize was worth it but always to be sure he got value for men's lives; know everything about administration but not do his staff's work for them; to be a master of decision but able to separate himself from those he commands to review the situation objectively; to be able to refresh his commanders and to know and be loved by his men; 'last, but indispensable, he must have that awareness for the enemy . . . that uncanny feel that sets a light

---

[4] Rhodes James, *op. cit.*
[5] He is the Aspinall-Oglander who wrote the excellent Official History cited in this book. His own name is mentioned only once in the work, and the part which Hamilton played in thought and execution seems to be well and dispassionately brought out.

winking in warning when something is brewing.'⁶ All these qualities Hamilton did have, but at Gallipoli they came to nothing because he would not enforce his will on those commanders to whom he had already set a task.

Haig, too, had most of the qualities that go to make a great commander. He certainly had that which is probably the easiest to acquire, a strategic knowledge of what it was good to do, of the direction in which a successful blow would hurt the enemy most, but this was marred by a lack of the instinctive flair for doing the unexpected. He also had that which cannot be acquired but is a gift of character, strong moral courage and the ability to make up his own mind and to make his subordinates willingly obey him. This was the reason why he always retained the confidence of the army. In general it may be said that in the Great War the remote higher command and the staff were unpopular, but this unpopularity was never transferred in men's minds to Haig himself. Remote and inarticulate as he was, his integrity was known and he was trusted. But if Haig is compared to his great predecessors Marlborough and Wellington, something is seen to be lacking. They made their mark in history because they so completely understood the warfare of their day, and because their tactics so completely matched the weapons of their day. Haig does not seem to have grasped the problem of the tactical battle in the same way, but rather to have believed that if an attack was persisted in with sufficient determination it must succeed. Some reason for this may be traced to the military background of the British Army itself, the transition from a series of small wars against sometimes savage and always ill-equipped enemies, the study of the American Civil War, and the sharp lesson of the South African War. All these had shown how often victory had eluded the commander who failed to persist just when victory was within his grasp. Hence the teaching discussed in Chapter 2, culminating in the dictum on positions which must be carried ' cost what it may '.⁷ And there can be no doubt that this teaching is sound, there can be no greater fallacy than to suppose that battles can be won painlessly. However brilliant the plan, in the end the soldier must advance, cost what it may,

⁶ Hamilton, *op. cit.*, especially pp. 150-1.
⁷ *See* p. 38.

to destroy the enemy. Where the enemy is prepared to fight, as the Germans always were, this means bloody battle and heavy casualties.

It can be said that Haig's underlying attitude to the problems of war was right, and that in 1914 and 1915 he showed the attributes of a commander which justified his selection to succeed French. In planning for the Somme, his first great battle in independent command, he showed some effort to understand the problems of trench warfare and to solve them. Haig was a cavalryman who looked at the early stages of the battle as the process by which that arm would be given the opportunity to seize victory. He was captivated by Birch, and perhaps blinded by science, into thinking that the artillery could blast a way through to this end. He neglected to see that the real brunt of the battle must be borne by the infantry, and that more than brute courage would be necessary for it to fulfil its predominant part. With all the advantages of hindsight it is easy to see that the Somme was pressed too long, but in Haig's defence there is evidence that the Somme went near to breaking the spirit of the German Army, and more than any other battle was responsible for the final destruction of the German Army.[8]

However Haig is judged over the Somme, it cannot be said that he drew from it all the lessons he should have. Although many commanders and staff officers were studying the problems that faced them there was no special effort at G.H.Q. to make an objective and vital study of the special problems of trench warfare. These were, briefly, how to surprise, overrun, and penetrate a well-sited defence system some four miles deep, the front edge of which was only a short distance from one's own, protected by massive wire entanglements and covered by the flanking fire of machine-guns and a wall of fire from artillery and mortars of all calibres sited in depth.[9] The attempt to do this almost entirely by artillery had forfeited surprise and had reduced the battlefield to a shambles and morass over which it was almost impossible to move.

Improved methods at Arras had pointed some of the way and

[8] *See* pp. 99-100.
[9] For this definition of the problem I am indebted to Major-General R. C. Money who served throughout the war on the western front in ranks from Subaltern to Major.

had brought out some of the problems that had to be solved. It showed above all the command problem, the difficulty in communication which prevented reserves being used at the right time and in the right place. In criticizing commanders for being out of touch with what was happening in front and of being tied to the rear echelons, the conditions of battle must be remembered. Means of communication were extremely unreliable; there were no light wireless sets suitable for use in forward areas, wire was cut by shell fire and tanks, visual light signals were crude and easily misread. Often the only method of getting a message through was by runner, or where tracks existed, despatch rider. In the forward areas this meant a man on foot, travelling in difficult ground at two miles per hour with the strong probability that he would be killed on the way. Often it cost several men's lives to get a message through. For a commanding officer to visit his forward units meant a dangerous foray lasting several hours. During all that time he was out of reach of any orders from his brigade commander, and so on up the chain of command. It is small wonder that each commander tied down the position of his next forward echelon of command. This is the kind of problem that ought to have been scientifically studied. Haig must bear the responsibility that it was not. It was almost insoluble, but the invention of the tank did offer possibilities. Wireless had been in use between aircraft and artillery since 1914[10] and it ought to have been possible to devise a tracked vehicle to carry a wireless set. A few mobile headquarter units of this type would have been invaluable for commanders of reserve formations and for an army commander fighting the battle. Such a solution might have come out of an objective study.

Haig was not unreceptive to the idea of the tank but he can hardly be said to have grasped its potentialities with both hands. Single minded enthusiasts like Swinton and Fuller are difficult to deal with, but Haig went too far in allowing his staff to restrain them. Haig can be criticized for the premature use of tanks but such criticism must be tempered by the knowledge that the tanks and tank crews that played such a part in 1918

[10] It was first used at the Battle of the Aisne by Lewis and James, two officers of 4th Squadron R.F.C. who were wireless enthusiasts. Both were killed.

could hardly have been produced without some at least of the experiences of 1916 and 1917. Fuller himself says: ' It must be remembered that, whatever tests are carried out under peace time conditions, the only true test of efficiency is war, consequently the final test a machine should get is its first battle; and until this test has been undergone no guarantee can be given of its real worth and no certain deductions can be made as to its future improvements.'[11] But Haig can be blamed for not sufficiently directing and driving his army commanders and his staff towards evolving a proper technique for the use of tanks. Haig regarded tanks merely as an adjunct to infantry; this is all the more reason why he should have insisted on the working out of a proper system of co-operation between them. Until the preparations for Cambrai there was no combined training and even then the drill worked out covered only the earlier stages of the attack, always the part of the battle that caused least difficulty. It was not until men like Maxse and Monash worked on the problem that a really efficient system of training and fighting was evolved. To achieve that, as has been seen, Monash needed a good deal of pressing to use the tanks at all.

Obviously a man with Haig's awful responsibility could not be expected to work on these problems of minor tactics, but they were the essentials of victory and they were problems on which the Germans always had a team working. There is no doubt that Haig's action in appointing Maxse with a special staff to study questions of training and tactics came two years later than it should have done. Credit must be given to the progress in artillery technique that went on throughout the war; it is a pity that the same professional approach was not made to the infantry problems.

Something was done to improve the standard of training in 1917 and in general the attacks in the Third Ypres series of operations were better carried out than those on the Somme. Yet the general pattern of tactics was the same and most of the attacks were persisted in far beyond the hope of success. After all that had happened on the Somme it is difficult to forgive the launching of the main offensive at Third Ypres without any attempt to achieve surprise. There was no reason to

[11] Fuller, *Tanks in the Great War* (1920), p. 59.

believe that the preponderance in artillery and the methods which achieved little on the Somme would do more at Ypres. In Haig's defence it may be said that what Haig predicted in 1916 and 1917 he predicted again in 1918; he was the only one to do so and this time he achieved his aim. There is little doubt that at this stage Haig's vision and persistence shortened the war by at least six months. But a study of the 1918 operations shows that they succeeded because the British had learnt from Cambrai and from the German spring offensive much that they ought to have learnt earlier.

The extent to which Haig persisted in the offensive in 1916 and 1917 in order to save the French has been the subject of endless controversy. The fiftieth anniversary of Passchendaele brought it out anew and it is doubtful if it will be ever resolved. There is no doubt that Verdun in 1916 and the mutinies following the failure of the Nivelle offensive in 1917 did put a responsibility on the British Army which could only be fully fulfilled by successful offensives. On the other hand Haig himself did not put forward this reason for his persistence until ten years after the war. He persisted because he knew that the war could be won only by the defeat of the German Army in battle and he believed that a series of blows, costly as they were, would achieve that end. The idea that Haig was never a free agent seems to have come rather from Edmonds, the official historian, than from Haig himself. Edmonds was a great admirer of Haig and it is not unlikely that, seeing the pattern of the war as a whole in retrospect, he worked out in his own mind this explanation of Haig's persistence in the face of mounting casualties. It seems more likely that the persistence, whether justified or not, was the natural outcome of Haig's character and qualities. He felt, perhaps, that his own inflexibility of will had only to be matched by those below him to achieve success. At the height of the Battle of the Somme he wrote:[12]

> Proof given to the world that the Allies are capable of making and maintaining a vigorous offensive and of driving the enemy's best troops from the strongest positions has

[12] *Private Papers of Douglas Haig*, ed. Blake (Eyre & Spottiswoode, 1952), pp. 157-8.

shaken the faith of Germans, of their friends, of doubting neutrals in the invincibility of Germany. Also impressed on the world England's strength and determination and fighting powers of the British race.

This last sentence and another note in his diary, ' Principle on which we should act " maintain the offensive " ' show in some degree the blindness of his own determination and that which he expected in others. It is certainly true that in war success can only be achieved by the ability of the man in the ranks to do what seems impossible, and that this ability depends upon leadership, morale, and discipline instilled from above. But Haig perhaps failed to see that a dead man cannot advance, and that to replace him is only to provide another corpse.

The two outstanding military writers after the war are scathing in their criticism of the conduct of operations on the western front. Fuller wrote of Haig's ' inability to grasp what appeared so obvious '. He also wrote:

It is indeed strange that the man whose stubbornness in the offensive had all but ruined us on the Somme, should from August (1918) onwards have become the driving force of the Allied armies. Yet this was so and it must stand to his credit, for no man can deny that, during the last hundred days of the war he fitted events as a hand fitted a glove.[13]

Liddell Hart does something to see through the enigma when he writes:

Haig showed considerably more realization of the tactical problem than most of the heads of his staff or his army commanders, but was handicapped by being a cavalryman. He made some good suggestions before the Somme offensive, but deferred to the objections raised by Rawlinson and Montgomery-Massingberd. He also showed an early inclination in August to respond to Byng's proposal for an attack at Cambrai with massed tanks, but gave way to the orthodox objections of his chief of staff, Kiggell, and deferred trying until it was too late. He was also uneasy whenever he had

[13] Fuller, *op. cit.*, p. 341.

to deal with gunners, and felt that their technique was a mystery he was not able to question. That was why under the influence of Curly Birch he continued so long to accept the necessity of a very prolonged bombardment at the sacrifice of surprise.[14]

Allenby's success in Palestine may serve to some extent as a measure of the difficulties which faced Haig. Allenby, a cavalryman, had come near to success at Arras, but had failed because he could not solve those problems of the command and placing of reserves which had baffled Haig and others. He had gone on to make the same mistake of pressing the battle beyond hope of success. Given independent command in Palestine, a country where there was inevitably one open flank, he achieved brilliant success from his first battle, at Gaza and Beersheba, up to the final defeat of the Turks at Megiddo and the pursuit beyond. Here he had scope for imagination, for deception, for rehearsal and training, and for the use of that mobility which cavalry could give even in the face of the Turkish qualities in defence which commanders and men had displayed at Gallipoli. Another aspect of command shown by Allenby was his use of Lawrence and the Arabs. His relationship was a model of how a commander can use a subordinate whose very qualities and whose independent role make for difficulties.

[14] Letter to the author, 4 Feb. 1965.

Chapter 9

BETWEEN THE WARS

It is usually defeat that brings military lessons, while victory, as after Waterloo, brings complacence. After 1918 there was plenty of complacence for ' the war to end wars ' had been won and armed forces now hardly seemed necessary. On the other hand, on the military side there was great heart searching. Some of those who had held, or who now held, the highest positions submitted too easily to the complacency and were unwilling to examine closely the reason why success had been so costly, but there was arising a generation of young commanders and staff officers who had seen too much that was wrong and muddled and they were determined that the army should learn from its mistakes.

Among the writings that affected military thought of the day two different classes of writer deserve notice. First Robertson and Maurice who wrote outstandingly good books on the higher direction of war and secondly Liddell Hart and Fuller, both soldiers who became professional writers and discussed the history and theory of war in all its aspects. Robertson did the greatest service to the nation and to the army by writing *Soldiers and Statesmen,* an account of his relationship as C.I.G.S. with the government and with the commanders in the field. The book covers the whole conflict of strategical opinion during the war. Robertson was quite certain that his ideas had been proved right in the end but the book was written dispassionately and made possible a just assessment of the facts. As C.I.G.S. Robertson had not only fulfilled his function as military head of the army but he had carried out almost all the functions later invested in the Chiefs of Staff Committee. It has been shown how he first had to get right the duties of the military head of the army which had got out of gear with the appointment of a soldier, Kitchener, as Secretary of State for War. He had also to regulate his position with the other military members of the Army Council to whom his relationship was not that of superior but of *primus inter pares*. There is no doubt that Robertson's book, so clearly setting out the problem, was invaluable in the later evolution of the Chiefs of Staff Committee, which was to fulfil its function so well in the second world war. It created just that knowledge and atmosphere in military and political circles which enabled Hankey to become the architect of a sound system. Hankey served successively as secretary to the Committee of Imperial Defence and the Chiefs of Staff Committee (1912-1938), the War Cabinet (1916-18), and the Cabinet (1919-38) and was able to hand over his mantle to Ismay, who had so often served as his deputy. It is part of Hankey's great service that he was always a secretary and never attempted to wield or assume power; he was therefore trusted by all and his great knowledge and experience was of inestimable value to successive heads of government and service chiefs in building up the defence machinery which was in being in 1939.

Maurice's contribution to post-war military study was less directly pointed at the lessons of the war, but its more general and theoretical aspect had no less an effect on the minds of those who were to shape our future strategy. He was the son of that Maurice who was one of Wolseley's most trusted subordinates. The elder Maurice has been described as ' The second pen of Sir Garnet ',[1] and he preceded Henderson as Professor of Military History at the Staff College. The younger Maurice was therefore steeped in military history, but he was also a practical soldier of wide experience. He had been head of the operations section of the B.E.F. and had gone with Robertson to the War Office to become Director of Military Operations. He had been obliged to vacate this appointment because in 1918 he became involved in public controversy. He felt so strongly about a statement by Lloyd George in the House of Commons he felt bound to intervene. Lloyd George stated that despite the losses at Passchendaele the British Army in France was stronger on 1 January 1918 than it had been on 1 January 1917; Maurice wrote to *The Times* giving the correct figures. The whole incident was unnecessary because it was the result of a misunderstanding on both sides. Wrong figures had been provided by the War Office and then amendments had

[1] Luvaas, *The Education of an Army* (Cassell, 1965), ch. 6.

been sent to the Prime Minister. Through a mistake in his office the correct figures had never reached him. The truth was brought to light by Lord Beaverbrook after the war.[2]

Maurice's character can be seen from a delightful description by Major-General Spears, written about a visit of Robertson and Maurice to France at the time of the Nivelle controversy.

As imperturbable as a fish, always unruffled, the sort of man who would eat porridge by gaslight in winter looking as if he had enjoyed a cold bath, all aglow with soap and water, just as cheerful as if he were eating a peach in a sunny garden in August . . . A little *distrait* owing to his great inner concentration, he simply demolished work, never forgot anything, knew everything, was quite impervious to the moods of his chief, the accurate interpreter of his grunts and groans, and his most efficient if not outwardly brilliant second. No man ever wasted fewer words or expressed himself when he spoke with greater clarity and conciseness. An admirable character, the soul of military honour, with a deep sense of civic duty inherited from a family which placed service to the country and to the people of the country above all else, he too suffered acutely from the tactics of the politicians and their too subtle methods . . . Robertson and Maurice were torn by conflicting loyalties. They did their best to serve their political chiefs faithfully, but they had a higher duty to the nation. They had to carry out the policy of the government, but it was incumbent on them to see that the policy was sound and likely to lead to victory.[3]

After the war Maurice became Professor of Military Studies at London University and wrote *British Strategy*[4] which was a classic of the pre-nuclear era. In this book he examined the principles of war as illustrated by history, to show what light they shed upon the British defence problem. It is a work which profoundly affected the outlook and helped to form the strategic thinking of the generation of soldiers who were to guide our destinies in the second world war. Another book

---

[2] Beaverbrook, *Men and Power* (Hutchinson, 1956), ch. 8.
[3] Spears, *op. cit.*, pp. 35-6.
[4] Published by Constable, 1929.

which contributed to the same end was *Imperial Military Geography* written by an officer of the Army Educational Corps, Captain (afterwards Brigadier) D. H. Cole.[5] This was an altogether different book and much more elementary in its approach. Nevertheless it contained a thorough and scholarly analysis of the facts on which the strategy of a great maritime power must be based. It had its effect more on the junior officers whose military minds were being formed in these years and whose military contribution on the planning, operational, and War Office staffs was to be so important. There can be few officers who took the promotion and staff college examinations —certainly none who passed—who did not reap benefit from this admirable book.

Fuller and Liddell Hart were as interested in the tactics and technique of war as in strategy. Independently they produced far-sighted and original treatises on mechanized warfare and, as several German generals have acknowledged, evolved the theories on which the Germans were to organize their *Panzer* forces and develop their *Blitzkrieg*. Fuller was a serving officer up to 1933 and some of his work as such will be discussed below; some of his best and most provocative works were written before he retired. In a prize essay (published in the *R.U.S.I. Journal* in May 1920 under the title ' The Application of Recent Developments in Mechanics and other Scientific Knowledge to Preparation and Training for Future War on Land.') he for the first time codified the principles of war. Many soldiers and historians, Napoleon, Clausewitz, Jomini, and Foch among them, had referred to the principles of war but none had ever tried to list them or to express them in comprehensive and concise form. The principles enunciated by Fuller were included later in the first post-war edition of Field Service Regulations and have remained virtually unchanged in our military literature since.

Liddell Hart used his deep study of military history to develop the theory of the indirect approach, and to instil into his readers the idea that the longest way round was often the shortest way there. He insisted that the object of manoeuvre and of battle was to disrupt the mind of the opposing com-

---

[5] Published by Siften Praed, 1924, and revised almost annually up to 1939.

mander, and that this could hardly be achieved without surprise. He also developed the theory of infiltration by writing of what he called the ' expanding torrent '.[6] Fuller's writing was deep, analytical and often abstruse and philosophical; Liddell Hart's work was easier meat for the younger officer, but was no less the product of deep thought and study.[7]

Both Fuller and Liddell Hart met with a good deal of opposition from the old school, and from some of the younger school too. The innate conservatism of man, and especially of men in established positions, was the principal reason, but some of the military and economic reasons for opposition were based on the government policy of the day and are discussed below. There is no doubt, however, that a strong added reason was that the adoption of their ideas would have meant the disappearance of the horse from the army. It was not that officers were dishonest in their opinions; there was plenty of evidence of the value of the horse. There is no doubt that the horse is better and quicker across country than man or machine, and there is no doubt that the hunting field gives a better opportunity than any other peace-time training for developing the eye for country, the initiative, and the dash so necessary for junior leaders in war. From these two premises it was easy to build, and arguments could be adduced pointing to the fact that the outstanding exercise of mobility in the field had been in Palestine where cavalry and mounted infantry had been most effective against the Turks, proved in defence as that formidable enemy was. Where the wish is father to the thought, a little evidence is enough to convince. The tank had proved itself too, but not in mobile operations so much as an adjunct to the prepared infantry and artillery assault. That even among devotees of the tank the merits of the horse were not overlooked is shown by a demand in 1928 that six horses should be issued to each tank battalion. It was considered that only thus could the requisite mobility for umpires in field exercises be ensured. The finance officers at the War Office did not see fit to accept the strongly supported recommendation, but horses were

---

[6] Liddell Hart, *The Framework of a Science of Infantry Tactics* (Hugh Rees, 1921), pp. 19-22.
[7] For a comprehensive review of the books and teaching of both authors see Luvaas, *op. cit.*, chs. 10, 11.

allowed for umpiring, and most usefully employed, in infantry brigades and battalions up to 1939.

Despite the opposition, both Fuller and Liddell Hart did have a real influence on the younger generation that was coming to the top. Their books were read and the arguments for and against their views were a frequent subject of articles in the principal military journals, the *R.U.S.I. Journal*, the *Army Quarterly*, and *Fighting Forces*. A criticism sometimes levelled at Liddell Hart was that he used history to illustrate his preconceived ideas rather than studied history to find the truth. In one of the letters of a fascinating correspondence between himself and Wavell in the 1930s Liddell Hart wrote:

> Your picture of me as a writer . . . does not recognize the change that has taken place gradually—one that I am very conscious of myself . . . My enthusiasm for promoting a cause has been superseded by a sheerly scientific interest in exploring facts to see where they lead without caring where that may be, as long as it is towards truth.[8]

Anyone who follows Liddell Hart's writings will probably agree with his own contention. Indeed the changes in his opinions which this search for truth involved led him just before the beginning of the second world war into what was probably his period of least influence with soldiers who had been brought up on his writings. He seemed at this time to suggest that because of our appeasement policy there was no longer any hope that France and Britain could frustrate Germany's designs in Europe; British involvement should therefore be limited and should be largely concentrated on defence against air attack. Liddell Hart's own *Memoirs*[9] throw some light on this phase, and show the effect of editorial control of *The Times* on the presentation of his views. Nevertheless his book *The Defence of Britain*[10] does not altogether dispel the impression felt at the time.

If we make a very general appreciation of the views put forward by the two writers we undoubtedly owe them a great

---

[8] Liddell Hart to Wavell, 30 July 1936.
[9] Published by Cassell, 1965.
[10] Published by Faber, 1939.

debt of gratitude for making the generation rising to command think out the lessons of the war. The most serious criticism that can be levelled at their teaching is that it under-estimated the value of the unarmoured anti-tank gun as the antidote to the tank. We shall see in a later chapter how Rommel was to develop the tactics of the gun to our disadvantage. Their theories might also be said to have lent weight to the suggestion that infantry and artillery as we had used them in 1918 were no longer necessary, and that all that was required was an armoured force containing an element of light infantry and self-propelled guns. No doubt both authors would be able to say that they were writing in general terms of the changes which the internal combustion engine had brought to war, and could point to writings that should have led us to see these changes for ourselves.

The influence of the Staff Colleges increased immeasurably after 1918. Officers selected for the directing staff and as students for the earliest post-war courses were all men who had proved themselves in war and in battle. When the entrance examination was resumed in 1921, it demanded a happy blend of theoretical knowledge and of past experience and ability to face the military problems of the day. The examination syllabus drawn up in those days has served without extensive modification to the present day. Indeed such changes as have been made have for the most part been to reduce the scope because, in these days when administration is our master, officers cannot be expected to devote so much time to study as in the more leisured days between the wars.[11] After 1918 there remained a few officers who considered that an officer who went to the Staff College was deserting his regiment, but they were fighting a losing battle and it soon became apparent that almost every officer who took his profession seriously tried to get to the Staff College. What was perhaps more important was that they were, for the most part, the best regimental officers. Apart from any achievements of the Staff Colleges themselves there is no doubt that the Staff College examination itself, with all the work it demanded from the officer in his late

---

[11] Far reaching changes in the educational structure in the Services now under consideration may involve the first major change in the Staff College examination system.

twenties and early thirties, was a most wholesome influence in the education of the army officer. It may fairly be claimed that the army set an example of adult education which did not exist then in any other profession. As an example of the attitude of the best officers who returned from the war, the author may cite his own regiment. The senior officers had all had the experience of going to war in 1914 as part of a well-trained unit which had quickly proved itself in battle. But they were conscious that they had not been taught to look deeply into the problems of war and to think out the new methods that would be necessary. They were determined that they would bring up a generation which were not only good regimental officers but would also know something of the business of war. Every officer that became adjutant between 1923 and the beginning of the war, except one, reached general's rank.[12] The exception, after distinguished service in command of a brigade in North West Europe, retired and became the editor of the *Army Quarterly*. In all, from those who served in this regiment between the wars came three generals, three lieutenant-generals, six major-generals, and nineteen brigadiers.

Scrutiny of the list of the Directing Staff at the Staff Colleges between the wars shows many of the names that were to become famous in the second war. Among the instructors at Camberley were Alanbrooke, O'Connor, and Slim; at Quetta, Auchinleck, and at each in turn, Montgomery. The successive commandants at Camberley were Anderson, Ironside, Gwynn, Dill, Gort, Armitage, Adam, Paget. Each of the regular officers that formed part of the Army Council which was unchanged throughout the years of preparation for and during the Battle for North-West Europe, Alanbrooke, Nye, Adam, and Riddell-Webster, had been an instructor at the Staff College. Fuller, too, was an instructor at Camberley in this period. He was too certain of his own deeply thought out theories and too intolerant of those who did not see the light so clearly as himself to be always acceptable, but his wrath was directed more at his superiors than at his pupils. He was an outstanding success as an instructor.

[12] 1st Battalion The Cameronians (Scottish Rifles). The adjutants were O'Connor, Galloway, Graham, Barclay, Murray, the author, Collingwood, Alexander. The other generals who served in the regiment in this period were: Riddell-Webster, Evetts, Money, Haugh and Frost.

Up to 1930 all officers who went to the Staff College had war service; the system of selection, which was partly by competition in the examination and partly by nomination of those who had qualified in the examination, allowed the officer's war record to count. After that date record was more difficult to assess, but the system of nomination still continued. It must be remembered that in those days an officer who sat the Staff College examination usually had some ten to twelve years regimental experience under some continuity of commanders both at regimental and higher level. There were some who said that the examination brought the wrong type to the top, but the war showed such fears to be groundless. Occasionally an officer with a brilliant staff record failed, and one or two successful divisional commanders had not been to the Staff College but the second war proved our system of officer training to be sound. In particular the practice of regular interchange between staff and command was healthy and a corner-stone of an efficient military edifice.

To turn now from the theoretical study of war to the more practical problem of the maintenance and training of the postwar army, the first point to remember is that life in Britain in these days was dominated by two factors, the hope of disarmament and a period of financial retrenchment manifested in the economic slump. The government felt that it could not afford and there was no compulsion to provide for an army fit to fight on the Continent. The basis of military policy was that there would be no war for ten years. This publicly declared assumption was relatively harmless when it was made in 1919, but the absurd fact was that it was automatically carried forward from year to year, was reaffirmed in 1929, and was not rescinded until 1932. The army in these years existed primarily for Imperial Policing and, as was inevitable in times of financial retrenchment, almost the whole of the military budget went on fixed charges such as pay, pensions, feeding, clothing, and housing of the forces.

Parsimony showed in other ways as well as the general organization of the army. Despite poor recruiting, few incentives either in pay or amenities were offered. In 1925 pay actually went down. A man who joined the infantry for seven years with the colours and five with the reserve (other arms varied slightly) would spend little more than the first year at home; he would spend the remaining six years, plus most of the additional year for which he could legally be held, overseas. During this time he would have no leave at home and there were almost no facilities for leave abroad. Apart from the fact that the soldier did not like such long unbroken periods of service overseas, this had a most serious effect on the army at home. Only abroad were units kept up to effective strength and manned by experienced soldiers. Units at home were not only grossly under strength, but were gravely disrupted every autumn to provide drafts for the foreign service battalion. There was more continuity for officers. Officers and those men who had extended their service beyond seven years were entitled to transfer to home service after six years continuously overseas. For officers, however, service abroad was much preferable to service at home. Abroad he had responsibility for a full strength unit and the whole life and well-being of his men were centred on the battalion or regiment. Moreover outdoor pursuits were much cheaper than at home, so that an officer, even if he had no private means could afford to shoot and fish and hunt, even to play polo; in fact to live the healthy life of a country gentleman. At his own expense the officer could take leave at home once in three years, and most officers did so. The greater part of the army abroad was stationed in India and formed part of the homogeneous Army in India, in which the proportion was about one British to three Indian units. The Army in India was subject to the government of India and not directly to the British government. The result was that the government of India had a considerable say in the shape of British units and in their training; all activities in India were directed to Imperial Policing, either active operations on the frontier or internal security in the big cities. Mechanization, therefore, lagged far behind and training for a major war very much took second place. On the other hand the opportunity for developing initiative in junior leaders and of gaining some experience of active service was much greater than elsewhere.

Relationship between the officers of the British and the Indian armies was good. Commanders and staff officers from both services worked side by side in the same headquarters. Among the British service officers who held important appointments in India between the wars are Rawlinson, Chetwode, Gort, Alexander, Montgomery, Paget, O'Connor, Morgan, Gale, and Gott. Among Indian Army officers recognized as being the best of their generation were Auchinleck, Slim, Gracey, and Messervy. There was a good deal of banter between the two services, the British Army would talk of ' the sloth belt ' and ' the *bail gari* (ox-wagon) mind ', and the Indian Army would point out that a British unit could not move without an immense administrative train and large canteen facilities; but on the whole there was friendship and understanding between the two. This was just as well since Indian divisions were to make a valuable contribution in the war in the Desert and to play a major part in the war in Burma.

At home, as was to be expected, great use was made in post-war training of the experience and studies of Maxse and his team in the Inspectorate of Training. Maxse was too senior to be Director of Military Training, a major-general's appointment in the War Office, but as General Officer Commanding in Chief at Northern Command from 1919 to 1923 he set the pace and introduced a considerable number of innovations in training. To his team of assistants were now added Liddell Hart, who joined Dugan's staff, and Montgomery as a major on his own staff. Maxse was responsible for the article on ' Infantry ' which appeared in the Twelfth edition of the *Encyclopaedia Britannica*. He was not so interested in producing polished prose as in the ideas and the drive in putting them across, and he got Evetts and Liddell Hart to write the article, giving them both the credit and the fee.[13] C. F. Atkinson, the military editor, described this article as ' the best thing in fundamentals that has been put on paper since the war. There is nothing like it here, nor in France, may I say.'[14] Maxse also kept a close watch on the production of *Infantry Training 1921* which, at his instigation, Dugan and Liddell Hart were preparing.

It was, however, not on writing but on training methods that Maxse was using his energy. He did much to introduce a system of minor tactics which could be taught as drills. In this he came up against some opposition from those who regarded

[13] Maxse Papers.
[14] Letter to Maxse, 6 Sept. 1921. *Ibid.*

close-order drill on the barrack square as something to be divorced from field training. It is one of the blind spots even of a number of enlightened officers that they forget that close order drill was the battle drill of Marlborough and of Wellington and the source of their victories in battle. One of our great military failures has been the slowness in the introduction of an infantry battle drill. Perhaps the reason why the Royal Artillery is our only truly professional fighting arm is that their barrack square drill is the drill with which they serve their guns in battle. Maxse did devise an excellent system of teaching minor tactics which was embodied in a small pamphlet entitled *Platoon Training* issued by the War Office in 1919.[15] This is incomparably the best book on minor tactics so far issued, far better than the manuals issued in the hey-day of battle drill in Home Forces under Paget in 1942.

Another training reform initiated by Maxse was the cadre system of training junior leaders. He was fond of catch phrases which sound trite to our ears, but which were effective. Here the phrase was ' teach the trainer how to teach before you teach the tommy '. He required each Depot or unit to retain a special cadre of instructors which would be available to train junior N.C.Os. The Depot cadre was also made available to help Territorial Army units. The whole system of individual training in use in peace and war since that time stems from the reforms instituted by Maxse while he was at Northern Command.

The artillery also went ahead to develop their technique of mobile war based on their war experiences culminating in the Battle of Amiens 1918. The most important manifestation of this was the issue of a training pamphlet in 1934 entitled ' The Fire Plan in the Infantry Brigade Battle '. Ideas which sprang from studies by the staff of the School of Artillery were here developed into a procedure for reconnaissance, making an outline plan, dovetailing the requirements of the various units, making a plan using to the best advantage all the weapons available, and the final issue of orders. The system was extremely flexible and enabled the fire of the whole divisional artillery and additional artillery to be concentrated on one part of the front or quickly switched to the point where it was

[15] The author still has his copy issued at Sandhurst in 1923.

needed. This procedure became the practice for orders, reconnaissance and committing units to battle which was in use during practically the whole of the second war. It was, indeed, the failure to remember these lessons which led to the series of defeats we suffered in the Desert in 1942.

Milne had become C.I.G.S. in 1926 full of reforming zeal. He had Fuller, fresh from three years as an instructor at the Staff College, as his military assistant, and began at once to set about the reform of the army. He was, however, soon to find out that the C.I.G.S. could not do as he liked. He was no more than the first member of the Army Council, *primus inter pares* as far as his other three military colleagues were concerned and subject to the strict financial control of the Permanent Under Secretary on the authority of the Secretary of State. Despite Milne's retention of the appointment for seven years in place of the normal four, he never really directed the War Office or the army towards the reforms he had intended. The effect of financial stringency on the development of tanks and weapons is discussed below, but financial control worked in other ways also in preventing the organization of a modern army. Changes in the army at home had to be achieved without increasing its size or cost and without unduly affecting its ability to fulfil its primary role of supplying drafts and reinforcements for overseas garrisons, of which the army in India formed the greater part. Any considerable increase in mechanized forces would require the conversion of cavalry to provide the new armoured units. India, however, refused to accept mechanized in place of horsed units on the grounds that they were unsuitable to the country and unduly expensive. The balance between our home and overseas army, always a central factor in our organization, would therefore be upset. There is no doubt that if the Army Council had been wholehearted in their conviction that mechanization was necessary and had been able to convince the government that preparations for a major war were necessary, these very real difficulties could have been overcome. Inter-service rivalries came into the question too, since the Army Council were faced with the argument that any money available for modernization should be spent on the Air Force and on anti-aircraft defence, and also with the view that the R.A.F. could take over some of the Imperial Policing duties from the army and so reduce the size of the army at home and overseas.

Milne was responsible for initiating one most important study when he convened a War Office committee to make a special study of the lessons of the Great War. This committee, presided over by Lieutenant-General Kirke with five major-generals and two brigadiers as members, produced a report which had a considerable influence on the training of the army. By the time the committee had completed its work Montgomery Massingberd had become C.I.G.S. and he restricted the issue of the full report to officers holding the most senior appointments in the army. He did, however, issue an excellent abridged edition down to headquarters of company, squadron, and battery to serve as a basis for thought, study, and discussion by officers. The members of the Kirke Committee were all forward-looking officers, well-versed in the views of the reforming element; a number of them were on close terms with Liddell Hart and were in the habit of discussing his theories with him. The abridged edition was a most valuable document; to read it in the light of the Second World War is to realize how enlightened were the members of the committee and how well they understood the lessons of the past. Emphasis was laid on three points; the importance of a continuous study of scientific developments and the necessity for the co-operation of civilian experts in peace time, the overriding importance of surprise whether in attack or defence, and the problem of maintaining the momentum of the offensive. In seeking a solution of the last problem the study pointed to the intelligent use of artillery, the use of armoured fighting vehicles (of which it said the Great War had placed the value beyond all doubt), and the training of infantry, better armed, carrying a lighter load, and capable of infiltration. The importance of night operations was also stressed and the doubt expressed whether, since the war, sufficient training in this direction was carried out.

Financial stringency had its greatest effect in limiting the production of new equipment and thus on the progress of mechanization and the development of armoured forces. The first step to build on the experience of 1918 and to experiment with the type of force then foreshadowed was the establishment in 1927 of 'The Experimental Force' based on 7th Infantry Brigade,

Plate 8. Tanks in action on the western front in October 1918. It is interesting to note the formation in depth and absence of waves of infantry in line.

Plate 9. Tanks, each accompanied by a group of infantry, passing through the main street of a recently captured village near Caumont, July 1944.

Plate 10. The Battle of Vimy Ridge, 12 April 1917. Tanks and troops advancing over the country. The leading infantry using the line formation, later to be discarded.

part of 3rd Division on Salisbury Plain. It had originally been intended that Fuller should command this force. When he discovered that command of the experimental force was to be combined with command of an infantry brigade and with the Tidworth garrison, and that the armoured forces would only be under his command for experimental training, he refused the appointment. Although in theory Fuller was right, this is an example of the somewhat truculent attitude which prevented his fulfilling all the tasks for which his great genius fitted him. The British way of doing things by compromise and evolution would have worked here admirably. If, instead of arguing for all that he believed necessary, he had accepted the command and used his good sense as to the amount of routine work he could delegate, there is no doubt he could have achieved much of what he set out to do. He and the commander of 3rd Division, Burnett-Stuart, would have made a magnificent combination.[16]

Fuller's attitude was typical of many of the high apostles of the tank. They suffered much from the slowness of more conservative minds to see the potentialities of the tank and the future of armoured warfare. Perhaps it was this that made them impatient of other people's ideas, but certain it is that they thought that only armoured forces counted, and that all who did not think exactly as they did were sinning against the light in deliberate obstruction of progress. Fuller, Hobart, and Martel were among the men of brilliant intellect who were exponents of the tank; none of them could be said to be easy to deal with except on exactly their own terms. The official view was that tanks were most useful as an adjunct to infantry, as they had been in 1918. The exponents of armour saw future warfare as a kind of naval warfare consisting of tank versus tank action, and wanted mobile formations based on the tank and almost entirely armoured. It is true that the Germans worked on this idea and achieved great success in Poland in 1939 and France in 1940, but there they were employed against armies in no way organized or equipped for modern battle, fighting in the face of almost complete German air supremacy and they were followed by a large number of normal divisions. Poland was, moreover, almost impossible, strategically, to

---

[16] A detailed account of the incident is contained in Liddell Hart's *Memoirs*, vol. I (Cassell, 1965), ch. 5.

defend. Despite their successes the Germans learnt much about armoured warfare in these campaigns and later, in the Desert, we are to see how impotent tanks were against a force of all arms, properly handled by Rommel. On the British side O'Connor had shown the same skill in handling a well-balanced force.

It was not only in command and tactical thought that the British failed to reap the fruit from brilliant ideas. Although they had invented the tank, by the time of the Munich crisis in 1938 they still had not settled on the type of tank or tanks they wanted, and had no plans for production in quantity. In 1939 our first contingent of the B.E.F. included only one under-strength tank brigade. This consisted of only two battalions, one equipped with Mark I infantry tanks and the other with half Mark I and half Matildas. Both were heavily armoured slow-moving tanks, the Mark I was armed only with a machine-gun, the Matilda had also a two pounder gun for use against armour but with no capacity for firing high-explosive shells. The only British armoured division was not ready before Dunkirk but arrived, in a sadly incomplete state, in time to fight with the French in their hopeless attempt to halt the Germans on the Somme. The division had arrived in France with none of its normal infantry and artillery components. In addition to its two armoured brigades it had only a mixed anti-aircraft and anti-tank regiment, the former consisting only of Lewis guns and the latter of two pounders. In the brigades two-thirds of the tanks were light tanks, armed with the machine-gun, and one-third cruiser tanks which had only just been issued and were short of essential equipment. The reason for this failure to provide the weapons and equipment was partly financial, arising out of the deliberate policy not to prepare for a major war, but the soldiers cannot escape blame. Nothing was produced in quantity because there was always a better model or a better idea just round the corner. Never has there been a better example of the best being the enemy of the good. The specifications laid down by the military were almost always too elaborate—a failing which prevented our producing a first-class battle tank in quantity until the Centurion, well after the war was over. In peace this fault meant that the small amount of money available went even less far than it should have done, and mechanical reliability usually suffered. If we had approved a simple standard design and gradually improved it in successive ' marks ', there is little doubt that we should have had much more effective forces, and a greater capacity to expand production when the decision to prepare for a major war was taken. It is easy to be wise after the event and there are undoubtedly opposite examples of a failure to take an early decision on a major change. The most striking example is the refusal to switch anti-tank gun production to the six pounder on the grounds that it would slow down production of the two pounder.

In the light of after-events the outstanding failure in our tank design was the failure to foresee the need for a tank gun that would fire either armour-piercing or high-explosive shell. It was not until the arrival of the American Grant tank in the Desert in 1942 that we had a tank so armed. This failure stemmed directly from the idea of the tank versus tank battle and of modern warfare as an all-armoured affair. None of our tanks, not even the infantry tanks, had any effective means of dealing with the hidden or entrenched anti-tank gun. The omission is all the more strange in view of the fact that by 1929 our experimental training had shown the need for a number of close support tanks armed with mortars or howitzers capable of firing smoke or high-explosive shells. It should have been a short step from there to see that all tanks should have the capacity of firing both armour-piercing and high-explosive ammunition, preferably from the same gun.

On the credit side it must be remembered that we were the only European army to go to war with completely mechanized transport. The Germans, French, and Russians were all dependent on horse drawn transport.

Despite the financial stringency and a lack of direction in some high places, the army of the thirties was trained on sound lines and was capable of being developed into an efficient war machine. Officers of the calibre of Dill, Brooke, Wavell, Wilson, Alexander, Paget, Montgomery and O'Connor, and in the Indian Army Auchinleck and Slim, were coming to a seniority where they could and did exert considerable influence. The names of these officers who later excelled must not be taken to suggest that there were not many in the higher ranks and among their contemporaries who played a most effective part. The tragedy is that officers with a radical view of future warfare were not better used. As has been shown this was sometimes due to their own intolerance. As is often the case with reformers, they could not suffer fools gladly, and they regarded as fools, often wrongly, those who did not have exactly the same vision. The example of Fuller has already been quoted; Hobart, another brilliant tank commander, also found it difficult to work with his superiors. He trained 7th Armoured Division in Egypt which, as part of O'Connor's force, was to achieve so much in 1941. But his views were so strong and independent that he could not get on even with so reasonable and far sighted a man as Wavell, and he was relieved of command in 1939. He was later to come into his own again when selected by Montgomery to devise and organize all the specialized armoured units for the Normandy landing and the campaign in North West Europe.

From mid-1934 to early 1937 Montgomery, then a Colonel, was a senior instructor at the Staff College, Quetta.[17] During the whole of this time he was chief instructor to the junior division, in which officers spent the first of their two years there. In his memoirs he says little about this time, but there is no doubt that they were among the most formative years of his military career. He had just finished commanding a battalion and he had wide regimental and staff experience in peace and war, including a time as a junior instructor at the Staff College, Camberley. Now when he was master of what he could teach and had the time for thought and discussion almost free from administrative distractions, he could settle down and consolidate his tactical doctrine. The Commandant, Major-General Guy Williams, under whom he served, had just that blend of wise direction and easy control which allowed Montgomery to exert his best influence. Williams was, moreover, strong and knowledgeable on the wider and political aspects of war, fields in which Montgomery was less expert than in tactics. It was a perfect combination. To learn under Montgomery in those days was an inspiration, an inspiration that was to deepen when his former pupils found later that what he taught in peace about command in battle he practised

[17] This covered the two years in which the author was a student there.

in war. His impact on the students was interesting especially in one respect. He arrived in the middle of a year and the division which was in the middle of its first year, never took kindly to his teaching; he upset too many of their preconceived ideas and he seemed to them much too much of the schoolmaster. But he taught the two succeeding divisions from the beginning and they regarded him in a very different light. Some found his individual and forthright methods of instruction easier to take than others, but to a man they acknowledged him to be a master of his craft; above all they recognized his ability to see the problem clearly, without over-simplification. Montgomery got on well too with the junior directing staff of the division. They may be mentioned to show the diversity. Bruce was a single-minded officer of the Gurkhas who had climbed in one of the unsuccessful attempts on Everest and regarded most problems as requiring the same kind of determination. Messervy was a carefree officer of the Indian cavalry. He was a dashing rider to hounds and in the polo field. He was to show the same qualities in the Desert, not always with success, and was finally to be the corps commander in what was probably the most brilliantly conceived and executed battle of the war, Meiktila. De Fonblanque was a gunner and an international show jumper. In lectures he was a master of mixed metaphor but in outdoor exercises he was a master of exposition and criticism. His anathema was the selfish thruster, but his attitude can best be expressed in one of his own maxims ' a firm seat on an office stool cuts no ice with the troops '. Playfair was a sapper; he had not been a student at the staff college because he had been too busy writing an opera to take the examination. He was to be Director of Plans during some of the most difficult days in the war, then Major-General General Staff on Mountbatten's staff in South East Asia, and later official historian of the campaigns in the Middle East.

Montgomery used some of the contrivances and showmanship for which he was later noted. For example he wore strange headgear; in those days the khaki ' Bombay Bowler ' was universal for officers in plain clothes but he always wore the white round helmet which had been usual in 1913 and was now worn only by station-masters. His wife was still alive and he took part in the normal social life of the community; he was always austere in his habits but he was not a teetotaller. His foibles were less annoying to those who really knew him than to those who heard of them second-hand. Few who knew him at that time had any doubt that he was a remarkable instructor and that he had an understanding of war such as was given to few to possess. He had thought out his own methods of instruction and they are reminiscent of those used by Henderson almost fifty years before. He would take some battle, often a little known one, and deal with it on the large floor model made up to represent the battlefield. He would deal with the problem as it had faced the commanders in the past and would then deal with the same encounter with modern weapons. Among the battles so dealt with were Éthe in 1914, Magersfontein in 1899, and other operations in the Orange Free State in 1900. One of the lessons most often driven home was the influence of the commander on the battle, the necessity for his being where he could control and of not allowing the force to 'drift into battle', as he called it. He talked often of seeking 'the tactical advantage'. He was fond of such aphorisms; another he coined was, 'when the situation is vague keep concentrated and reconnoitre widely for information'. He also talked of the necessity for commanders who were 'mentally robust' and that was the quality he always looked for in those with whom he dealt. Montgomery had one great disappointment at Quetta. He had had a large share in the writing of the current edition of 'Infantry Training' but he felt he had a wider exposition of tactical doctrine to give the army. He therefore set out as a single thesis the substance of his tactical lectures and sent it to Gort, Director of Military Training in India, and to Armitage, Commandant of the Staff College, Camberley, suggesting that it should become the tactical doctrine for the army. Neither of them would accept it. Possibly the use of the word 'doctrine', a favourite with Montgomery in those days, frightened them and it was this rather than the ideas in the paper that they rejected. Time was to show that Montgomery was right.

In May 1937 Hore-Belisha became Secretary of State for War with the avowed purpose of reforming the army. The government did not yet admit that war was inevitable, but at least they were conscious of the possibility. Hore-Belisha was genuinely interested in the army, he had shown this by a tour of units overseas as long before as 1930. The arrival of an energetic and resourceful minister at the War Office gave the Army Council and the War Office staff the opportunity of doing much that they had long wanted to do. Much was done; the necessary steps were taken for improved equipment, including the introduction of bren carriers for the infantry; the Territorial Army was made more efficient and its size eventually doubled; the best barracks the army had ever had were those he got built; the promotion lists were overhauled and many good men brought to the top; in the junior ranks, too, quicker promotion was introduced and the deadening system by which an infantry officer might serve as a subaltern for fifteen or even twenty years was replaced. Hore-Belisha took Liddell Hart as his unofficial adviser and from that got much benefit in tackling the modernization of the equipment and training of the army. He was also able to discuss with Liddell Hart his choice of officers for promotion to the highest appointments. Yet it must be confessed that the army did not benefit from Hore-Belisha's regime nearly as much as it should have done. The trouble was that Hore-Belisha and the military hierarchy, even when he had put the men he wanted into it, could not work together. There were faults on both sides. Even before Liddell Hart became Hore-Belisha's unofficial adviser there was friction but later the basic difference was over the question of unofficial advice. There was nothing new about a Secretary of State getting unofficial advice. Esher is one example of an unofficial adviser with unbounded influence on Sovereign and ministers. Certainly a Secretary of State will not be merely a rubber stamp to his generals in selecting the C.I.G.S., the Army Councillors, and the Commanders-in-Chief. But the extent and the depth to which unofficial advice is used is important and it is certain that an air of distrust was created over this and over one or two trivial and unnecessary reforms. Reform and questions of promotion are bound to generate resentment and opposition among the reactionary element which is to be found in every profession, but the tragedy is that even the officers whom Hore-Belisha brought to the top mistrusted him. The very fact that a good man was selected would obviously mean that he would have ideas and policies of his own to put into action. Hore-Belisha not only wanted to select the man but wanted to tell him what to do. In this respect Liddell Hart is not blameless; he also expected Hore-Belisha and the new men to do what he thought best. The older men had to go, the whole intention of the reform was to bring younger men to the top and quicken promotion throughout the army. But they could have been got rid of more gracefully, and could have helped as they went. Deverell, the C.I.G.S., had been a progressive soldier but became stubborn when Hore-Belisha dealt with others behind his back.[18] Knox, the Adjutant General, had all the work of the department at his finger tips and had the real interests of the rank and file at heart. He could have done much but he was not prepared to do things which he believed, with some justification, would undermine the discipline and loyalty of the army. He was replaced peremptorily by a younger, capable officer who was at no time as effective an Adjutant General as Knox. Hore-Belisha sought unofficial advice from within the army too, and this had even more dangerous possibilities for the loyalty and well-being of the army. Liddell Hart shows the extent to which Gort influenced Hore-Belisha while he was Military Secretary.[19] The Military Secretary does have a special position towards the Secretary of State and constitutionally is the only soldier who has access to him direct rather than through the Army Council. When Gort became C.I.G.S. he made quite certain that there must be an end to the practice of seeking advice indiscriminately from serving officers. He saw rightly that a healthy system could exist only if authority and the power of promotion lay through the appointed commanders to the Army Council.

For the three major appointments Hore-Belisha chose excellent men; Gort first as C.I.G.S. then as Commander-in-Chief B.E.F.; Ironside to replace Gort as C.I.G.S. and Dill for the Aldershot Command and so on to the outbreak of war for I Corps. Dill had been selected for his appointment before Hore-Belisha came, but it seemed at the time and still seems that Hore-Belisha put these three men in the wrong places. Gort was slightly out of his class in both appointments. He was

[18] Apart from much unpublished evidence this is made clear by the *Ironside Diaries*, ed. Mcleod & Kelly (Constable, 1952), ch. 3.
[19] Liddell Hart, *op. cit.*, vol. II, ch. 2.

a magnificent fighting soldier and would have been admirable in command of I Corps. Ironside was fifty-nine when war broke out and was past his best. As C.I.G.S. his impetuous nature was out of place. He had, moreover, a rather irresponsible streak in him and one has only to read his diaries[20] to see that he was not the man to act as principal military adviser to a government at war. On the other hand in command of the B.E.F., subordinate as it had to be to the French command, Ironside with his vitality, vision, and overwhelming personality, was possibly the only man who might have galvanized the French into activity. This comment is made in the knowledge that for the extrication of the B.E.F. from the situation in which they eventually found themselves, there could hardly have been a better man than Gort. Dill with a great strategical sense, wide horizons and balanced judgment was then at the height of his powers and was the obvious choice for C.I.G.S.

[20] *Ibid.*

Chapter 10

THE STRATEGY OF THE SECOND WORLD WAR

As war with Germany became inevitable the British government was forced to make up its mind whether to send an army to fight again side by side with the French Army or to confine its activities to maintaining a reserve in Egypt to secure the Suez Canal and other Imperial communications. In fact there was practically nothing equipped and ready to do either. As late as January 1939 a senior War Office official was able to answer a request for expenditure in connection with the despatch of a force to France with the statement that it had so far not been decided to send even one division to the Continent. Nevertheless in June 1938 a new branch, General Staff (Plans), had been formed in the War Office under Colonel Hawes[1] to overhaul such plans as there were to despatch an expeditionary force overseas, wherever it might be sent. The difficulties of working in the absence of a firm government decision can be imagined, but progress was gradually made and in spring 1939 matters were made a little easier by the government decision to open staff talks with France. Even then the situation was vague because the decision was taken that in a war with Germany and Italy we should have to remain on the defensive against Germany while we would take every opportunity of knocking out Italy.[2] Diplomatic efforts were continued to try to keep Italy out of the war and in June 1939 there was still no decision whether we would send two or four divisions to the Continent. Hawes, guided by Military Intelligence, who throughout had accurately forecast the political crises, gambled on four divisions on 1 September and went ahead with plans to move a B.E.F. on that basis.[3]

[1] Major-General L. A. Hawes. He had been an instructor of the Senior Division at Staff College Quetta during the author's time.
[2] Butler, *Grand Strategy*, vol. II (H.M.S.O., 1957), p. 10.
[3] Letter to the author, May 1965

The Strategy of the Second World War

The outbreak of war thus found the British in a position to send to France an expeditionary force of four well-equipped divisions, soon to be augmented by a fifth, but, as the last chapter has shown, woefully short of armour. The doubling of the Territorial Army in March 1939 gave them a good distribution of trained men for expansion both of their field army and their air defence units, but equipment for both was critically deficient; moreover plans for industrial expansion lagged far behind their requirements. To an officer in one of the divisions which had completed mobilization and moved to France, both in faultless programmes,[4] it was a shock to learn some months later of the shortages which existed in units training at home in preparation for their move to France. The period of quiet on the western front, known later as the 'phoney war' gave the British time to complete their expansion of forces and by the time fighting broke out in May 1940 they had ten divisions and one under-strength tank brigade in France. Industrial expansion is an even slower process than military expansion, but it cannot be said that Chamberlain made real efforts to use the time to the full and to organize the country ruthlessly for war; in this respect Churchill was almost a lone voice in the Cabinet.

In the Middle East for the time being the British peace-time garrisons in Egypt and Palestine were, with their French allies, considered sufficient to deal with the Italians should they come in against them. Dominion forces from Australia and New Zealand were under orders to concentrate in the Middle East and the South African continent was also available for Imperial defence in Africa. The defence of the Far East had the lowest priority and it was considered of paramount importance to keep Japan from joining the Axis. In India an expansion of the army was undertaken so that in addition to its primary roles of defence of India and internal security, the Army in India could provide forces as a strategic reserve. Such forces would be particularly valuable in the Middle East and Far East and some Indian battalions would relieve British battalions in internal security so that they could join field formations.

[4] The author went overseas as Brigade Major 2nd Infantry Brigade in 1st Division.

In deciding on their strategy the experience of 1914 to 1918 weighed heavily on the British and the French. Britain knew that her strength would not be gathered for at least two years, while both General Staffs trusted in the strength of the defensive and preferred to let Germany exhaust herself before they took up the offensive on the western front. The result was that Germany, incapable in 1939 of a war on two fronts, was able to devour Poland without the British being able to lift a finger to prevent it. Britain's problem was in no way eased by Italy's delay in entering the war against them. The British were deprived of the opportunity of taking the type of action against her that was well within the capacity of Britain and France together, while their concentration of forces in Egypt and their moves through the Mediterranean still had to be guarded against a sudden attack by Italy.

The defence problem on the western front was complicated by the attitude of Belgium and Holland. The Maginot Line, in which the French, and the British too, put such faith, ended at the Belgian frontier. The French were unwilling to continue the defences lest the Belgians should take this as an indication that she was to be left to her own fate and so seek an accommodation with Germany. On the other hand the Belgians would not make any open arrangements with the French which might prejudice the hope of remaining neutral. The Allies decided, therefore, that the Maginot Line would be continued northwards only by a line of field defences with an anti-tank obstacle and occasional pill boxes. General Gamelin, who had been on Joffre's staff in 1914 and like him combined the appointments of Chief of Staff of the Army and Commander-in-Chief, later decided that if the Germans advanced into Belgium the French armies on the left would advance to the line of the River Dyle, covering Brussels and joining up with the Belgian Army covering Antwerp. Gamelin insisted that the Allies should on no account risk getting engaged in an encounter-battle and that he would only advance if it was clear he would get *temps utile* on the Dyle position. On this understanding the British agreed and put the B.E.F. under General Georges who was the Commander of the French North-East front comprising all the forces along the Franco-German frontier.

The Strategy of the Second World War

During the time of inactivity in France the Russo-Finnish war seemed to give some opportunity for activity against Germany. It was hoped that with the consent of the Norwegian government the Allies might send a force to Narvik and so close the only winter outlet for Swedish ore going to Germany, and might also establish bases in Southern Norway which would enable them to forestall a German attempt to seize the orefields or to use Norwegian ports and airfields for attacks on Atlantic shipping. In the event the Germans moved first and so far from the Allies getting an unopposed footing in Norway they were forestalled at Narvik, Trondheim, and Stavanger while the Germans also mounted an operation against Oslo and overran the whole of Denmark. The British, to whom the French had entrusted the major part of the operations, concentrated on the capture of Narvik and Trondheim. Narvik was taken, but before the Norwegian adventure was over the German attack through Belgium and Holland had begun and this was the final reason for the abandonment of the footing obtained with such difficulty in Narvik and Northern Norway. It cannot be denied that from an allied point of view the campaign had been mismanaged. One of the reasons was that the majority of the units were being carried in His Majesty's ships and for naval reasons the ships were called away, depositing the men and taking away much needed equipment.[5] The chief reason, however, was the allied miscalculation of the German resources and efficiency[6] and to this was added some rather disturbing interference by Churchill, First Lord of the Admiralty, with the Naval Commander-in-Chief of the operations.[7]

On 10 May, the day that the Allies began their advance to the River Dyle, the Germans had already seized crossings over the Albert Canal, the Belgian covering position. Gamelin did not get his *temps utile* but it was not on the Dyle position that the Battle of France was lost. British participation in the advance and the subsequent operations leading up to the evacuation through Dunkirk left little scope for Gort to

---

[5] There is an excellent short account of the campaign in Collier, *The Second World War* (Collins, 1967), pp. 95-119.
[6] *Grand Strategy*, II, p. 149.
[7] Roskill, *The War at Sea* (H.M.S.O., 1954), p. 202.

influence operations. Thirteen of the best French divisions, the nine British divisions[8] and two-thirds of the Allied strength in armour were on the sixty mile front north of Namur, while on the ninety miles from Namur to Longuyon twelve divisions, some of them of doubtful quality, faced the main German attack by forty-four divisions under Runstedt, of which the spearhead was two powerful assault groups totalling seven armoured and three mechanized divisions. The Germans quickly established bridgeheads over the Meuse and punched a large hole in the French centre between the Sambre and the Aisne and by 20 May advanced elements had reached the Channel coast north of the Somme. The Allies had substantial forces north and south of this break-through and the Germans now required time to bring forward their main forces, particularly to hold their long and vulnerable southern flank. The French hope lay in striking across the Meuse near Sedan, a move not unlike that made on the Marne in 1914. Georges had already frittered away the three armoured divisions which had been in reserve near Reims, but now Gamelin took a hand. He saw that there were many more troops than were necessary behind the strongly fortified line south of Longuyon. There is little doubt that Gamelin should have replaced Georges in command and preferably taken over direct command himself. Instead he issued an order to Georges to prepare a counter-attack in the greatest possible strength to cross the Meuse below Sedan. The French First Army and the B.E.F. were heavily engaged by the northern army group under Bock but Gamelin thought, rightly, that an effort should be made to support the main counter-attack by a thrust south from Douai.[9] Although the Germans had used their air forces effectively in direct support of Runstedt's attacks there was no general air offensive over France and the railways and roads were available to make Gamelin's plan feasible.[10] The tragedy was that a few hours after Gamelin had issued his order he was replaced by Weygand, aged seventy-four and summoned specially from Syria. Georges therefore disregarded Gamelin's order and the precious ten or twelve days available for counter-attack were lost while Weygand made up his mind what to do.

[8] One division was temporarily attached to the French Army on the Saar.
[9] Goutard, *The Battle of France* 1940 (Muller, 1958), pp. 102-208.
[10] *Ibid.*, p. 206.

The Strategy of the Second World War   191

One of the striking features of this period was how little the British knew of the French intentions and capacity. From the point of view of Intelligence there is no doubt that they knew more of the German Army than of the French. This strange situation presumably arose because the British were so much the smaller contributor that they did not press the French for detailed information. The only indication of the malaise from which the French Army, or at least part of it, was suffering came from the Duke of Windsor who was acting as Senior Liaison Officer with the French Armies[11] and who throughout the 'phoney' war sent in disturbing reports about the lack of zest and the neglect of training in French units. An example of lack of knowledge of French strength may be cited from events after the battle had begun to go wrong. There was on the Situation Map in the War Office an enormous sausage behind Paris marked 'Réserve Générale d'Armée'. An officer of the Liaison Mission who had just returned from G.Q.G. was asked what was there, he answered in Cavalry language, unprintable here, to the effect that there was nothing. In Paris Gamelin made the same confession to Churchill.[12] Throughout the operations, however, Gort and G.H.Q. were aware of the situation and every move of the enemy was quickly reported to the War Office.[13] One of the principal sources of information was a team of liaison officers which Gort had with French formations. But Gort was powerless to act except as part of a co-ordinated French plan and there was no sign of this. Churchill had by this time become Prime Minister and, as might be expected, there were suggestions from him that Gort might attack south with the greater part of his nine divisions. Such suggestions took no account of the situation on Gort's own front and Gort himself was clear that the real effort must come from the south and that the most he could do was to co-operate with two or three divisions in the direction of Cambrai. In order to stabilize his own front Gort did launch a small counter-attack near Arras on 21 May. He used part of

[11] He was not at G.Q.G. when there was a British Mission under Major-General Howard-Vyse.
[12] Goutard, *op. cit.*, p. 187.
[13] The author was at this time in the Operations branch of the War Office dealing with B.E.F.

50th and 5th Divisions and his one tank brigade. The attack had considerable local effect and Rommel, whose *Panzer* division bore the weight of the attack, believed that he had been faced with a strong armoured attack.[14] The strength of Gort's understanding of the situation was that he realized from the beginning that if no major counter-attack was initiated by the French the B.E.F. would be faced with the alternatives of destruction or of evacuation through Dunkirk. He did not let this affect the strong resistance with which his army faced the enemy but he did ensure that his force remained concentrated and firmly astride his newly improvised lines of communication to the channel and that none of it was frittered away on useless operations. The first news of the contemplated evacuation was received in the War Office on 19 May and on the 20th a conference was held at Dover at which representatives of the War Office and the Ministry of Shipping met Admiral Ramsay.[15] As a result of Gort's foresight, the robust defence of the B.E.F. and the magnificent effort of the Royal Navy and Royal Air Force, almost the whole of the B.E.F. was brought back to England between 26 May and 4 June. There were left in France only 51st (Highland) Division which had now come up from the Saar to the Somme, the weak and ill-equipped 1st Armoured Division, and the troops still on our original lines of communication to Cherbourg and St Nazaire. Although the loss of men was small the whole of the heavy equipment of the B.E.F. was lost and the British were faced once more with the task of equipping an army.

After the Dunkirk evacuation the British decided that they would as soon as possible build up their forces in France again through the western ports. The speed with which the Germans disposed of the Allied forces on the Somme showed that there was little chance of the French creating the situation in which this would be possible. Brooke was selected to command the B.E.F. on re-entry, but his visit to the Cherbourg peninsula and his meeting with Weygand showed him that there was no possibility of effective action. Weygand was still thinking in terms of linear defence which events had so clearly shown to be inadequate; there was no hope of the French being able to

---

[14] *Rommel Papers*, ed. Liddell Hart (Collins, 1953), pp. 32-33.
[15] Roskill, *op. cit.*, p. 212. The author was present at this conference.

Plate 11. British Infantry advance in North Africa at the double to the ridge beyond disabled enemy tanks. November 1942 (after Alamein). It is interesting to contrast this with the infantry line seen in plate 10.

Plate 12. Sherman tanks pass through infantry holding positions near Caen, June 1944. Again note the absence of the linear formation used in World War I.

The Strategy of the Second World War

produce the check followed by a blow against the German communications which alone could stabilize the position. Brooke reported to the government in no uncertain terms. There was some doubt whether Brooke was intended only as commander of the leading corps on re-entry and whether Gort was later to return to France. In fact the decision had been taken that Brooke would become Commander-in-Chief. Gort was indignant about this and asked the Army Council what he had done wrong. The Army Council were bound to admit that Gort's conduct of the operations leading to the evacuation had been faultless. Nevertheless they considered that Brooke was the man to command our principal army and he became Commander-in-Chief Home Forces and prepared to meet invasion.

The chief effort of the country was now devoted to the defence of Britain against air attack and against the German attempt at invasion which seemed the natural consequence of her victory in Europe. In September Churchill in a comprehensive review of future action, drawn up to guide government departments in munition production, said that only the Royal Air Force could win the war and that a supreme effort must be made to gain an overwhelming mastery of the air. In qualification he said, ' The Fighters are our salvation, but the Bombers alone provide the means of victory.' The task he saw for the bombers was ' to pulverize the entire industry and scientific structure on which the war effort and economic life of the enemy depend '.[16] In the days when Churchill's paper was being discussed the R.A.F. ensured that the Germans never achieved the degree of air superiority which they considered essential for the invasion and so the army's plans for the defence of the British Isles was never put to the test. The long invasion scare lasted from Dunkirk until the situation was entirely changed in May 1941 by the German invasion of Russia. The long period of readiness not only gave the British the opportunity for the re-equipment of their army, but also gave them the occasion for large scale exercises. Officers selected for high command thus had experience in command of large forces in mobile exercises such as they had never had before.

[16] Churchill, *The Second World War*, vol. II (Cassell, 1949), pp. 405-406.

Despite the concentration on home defence and the air effort against Germany the British did not forget their pre-war determination to do everything possible to deal with Italy as soon as she joined in the war against them. The situation in Africa was greatly altered for the worse. The British had counted on the French and between them they hedged in all the Italian colonies, while in the Mediterranean they had hopes of matching Italian naval and air strength. In May 1940 Wavell had only 36,000 men in Egypt and 27,000 in Palestine and local forces in the Sudan and East Africa.[17] With these forces he was responsible for the defence of Egypt, the Sudan, Kenya, Palestine and Iraq against forces which far outnumbered him in Libya, Eritrea, and Abyssinia. In addition Wavell had responsibility for any operations that might be necessary in the Balkans or against French forces in North Africa or Syria and for the support of Turkey if she entered the war. Threatened as he was on every side Wavell was master of the situation. Apart from the loss of British Somaliland in August 1940 the position was held everywhere and by a series of well-designed operations, often necessitating the switch of divisions from one theatre to another, Wavell gained security and the initiative in the Middle East. First came O'Connor's glorious campaign in Libya;[18] then the conquest of Eritrea and Abyssinia. Italian power in Africa was broken and the only foothold left to her was in Tripoli.

The situation was changed when Hitler decided he could not allow Italy to be driven out of Africa and at the same time embarked on operations for the occupation of the Balkans. At first it seemed that Greece, which had already shown herself a match for Italy, was the only country that would resist. However, the Yugoslav people refused to let their government fall in with Hitler's wishes, but their action was too late to be effective. The German invasion of Greece faced Britain with a most difficult strategic decision. O'Connor at that moment had Tobruk almost within his grasp, and Tobruk was as far as Wavell had intended to go. But less than a month later O'Connor had rounded up the whole Italian army

---

[17] Playfair, *The Mediterranean and the Middle East*, vol. I (H.M.S.O., 1954), p. 93.
[18] See Ch. 11.

in Cyrenaica and despite his stretched communications and the air hazards of an advance to Tripoli O'Connor believed he could get there. There were certainly not sufficient resources in the Middle East both for this advance and for an expedition to Greece. The decision was made to go to Greece and a force of four divisions and an armoured brigade was put under orders. O'Connor's infantry division, 6th Australian, was part of this force and his armoured division with its worn-out tracks and vehicles was pulled back to Egypt for refitting. All the political advice was in favour of intervention in Greece but this was not a political decision forced on the military against their will. Both Wavell and Dill agreed with the decision. Dill now C.I.G.S. had come out with Eden to Cairo on 19 February (a fortnight after O'Connor's final victory at Beda Fomm) and went on to Athens. Nothing was achieved in Greece; the Greeks found themselves partly unwilling and partly unable to withdraw to the Aliakmon line where effective defence might have been possible. With Greek consent the British force was withdrawn and, as at Dunkirk, the whole of the heavy equipment was lost. In Africa not only did the British miss the opportunity to take Tripoli but almost all the fruits of O'Connor's victory were lost and they were back on the Egyptian frontier with an isolated garrison holding Tobruk. Most grievous of all, O'Connor who had gone forward to see Neame, now in command of the battle, was by a fortuitous accident captured. If it were possible to look at a strategic problem in military isolation the decision to go to Greece would have to be regarded as wrong. And yet with all the advantages of hindsight it is not possible to say that Dill and Wavell were wrong to give the advice they did. Defeat impresses nobody but sometimes a refusal to take action does more harm even than defeat. The British had stood by while Poland was overrun, they had fought well in Belgium but had been unable to influence the battle. Greece had captured the imagination of the British people by her determined fighting against Italy and if the British Army had failed to lift a finger to assist her the nation might have felt that it had lost its soul. In the words used by Eden in reporting back to the government: ' It is of course a gamble to send forces to the mainland of Europe to fight Germans at this time. No one can give a guarantee of success . . . it is better to suffer with the Greeks than to make no attempt to help them.'[19]

The Chiefs of Staff Committee had been functioning as a sub-committee of the Committee of Imperial Defence since 1923, and on the outbreak of war had been used under the War Cabinet as the instrument for the higher conduct of the war. The Committee of Imperial Defence thus went into abeyance. When Churchill became Prime Minister in May 1940 he had reconstituted a smaller War Cabinet but at the same time he initiated the system by which the day to day direction of the war was carried out by the Defence Committee, which was in fact the Chiefs of Staff Committee with himself added as chairman. Churchill assumed the duties of Minister of Defence as well as Prime Minister and the Defence Committee had a small Secretariat with General Ismay as Chief Secretary and Colonel Hollis and Lieutenant-Colonel Jacob as assistants. Under the Chiefs of Staff worked the Joint Planning Staff; there were a number of levels but all worked in the same way in that each had as members one officer from each service and he had free access to his own Service Ministry and was responsible for day to day touch with it. Thus the three Directors of Plans who were at the head of the joint planning organization were not only collectively responsible for what they did but each was also responsible to his own Director of Operations in the War Office, Admiralty or Air Ministry.

Notwithstanding the functions of the Defence Committee and his own direct and dynamic interest in the war, Churchill never at any time or in any way disregarded the Cabinet or forgot its ultimate responsibility. He was careful to ensure that the Cabinet had its proper opportunity to decide constitutional and political questions and that ministers who might be concerned with a particular question should attend when it was discussed by the Defence Committee. The Chiefs of Staff were thus integrated into the government machinery and yet were free to give their whole mind to the conduct of the war. Churchill's direction and the use he made of his military advisers is one of the most fascinating aspects of the war. The lessons of the Great War had been well learned and the

[19] Quoted *Grand Strategy*, vol. II, p. 441.

The Strategy of the Second World War

machinery had been tailored under the guidance of Hankey[20] for the purpose for which it was now needed. But machinery can be neglected or ill-used, and it might have been thought that Churchill, with his strategic flair and his unbounded confidence in his ability to decide what it was best to do, might have disregarded his military advisers and come to a conflict like that between Lloyd George and Robertson. This did not happen. Churchill loved to issue long and detailed instructions to a Commander-in-Chief; that to Wavell in August 1940 is a good example.[21] But such instructions were first discussed and modified in the Defence Committee and in the example cited it was made clear that it was a directive for guidance and not a detailed order. There is no doubt, however, that at this time Churchill's opinion of Wavell and of Cunningham, the Naval Commander-in-Chief, was at odds with that of the Chiefs of Staff and that he much underestimated the two great commanders.[22] Dill who understood the ability, determination, and energy of Wavell, did not spend time in fruitless argument with Churchill but put it to him that he must either trust Wavell or sack him. Churchill did eventually replace him, but not before Wavell had used his scant forces to the best advantage and completely changed the situation in Africa.

Churchill's method with the Chiefs of Staff was to suggest, to challenge, to chide, sometimes to bully, but never to override. To serve him was an exhausting task, especially in those days when the British stood alone and were still gathering their strength. Dill who bore the brunt of these days was worn out by the time the Japanese attack in the Pacific brought the United States into the war. Churchill rightly judged that a change of C.I.G.S. was necessary and Brooke became C.I.G.S. and Chairman of the Chiefs of Staff Committee. Brooke's brilliant insight moved him to suggest that Churchill should make Dill his personal representative in Washington.[23] In that capacity Dill was able to meet almost daily with the United States Joint Chiefs of Staff so that together they constituted the Combined Chiefs of Staff in permanent session.

The relationship of Brooke to the Prime Minister brought

[20] See p. 164.
[21] Churchill, *op. cit.*, vol. II, pp. 379-382.
[22] *Grand Strategy*, vol. II, pp. 310-311.
[23] Bryant, *Turn of the Tide* (Collins, 1957), p. 284.

a new vitality to the Defence Committee and the Chiefs of Staff. The publication of much of the Alanbrooke Diaries[24] has, despite the warning in the author's two preludes, given to many a false suggestion of conflict between Brooke and the Prime Minister. The diaries may have revealed an inner tension which was eased by a daily blowing off of steam. None of this was apparent in Brooke's work in War Office or in committee. He was imperturbability itself and, in addition to his ability to grasp a complicated situation in a flash, he was a master of exposition. His relationship with Churchill was exactly what that between a Prime Minister and his principal military adviser ought to be. He had an admiration for Churchill and at the same time a clear idea both of his genius and his shortcomings. A short time after the war Brooke said of Churchill that he had never known a man with such a brilliant insight into broad strategy but who was at the same time so ignorant of the mechanics of war, that is to say of what a battalion or a division could accomplish.[25] This judgment reflects the difficulties under which Brooke and the other Chiefs of Staff worked and also the opportunities which were offered by Churchill's remarkable leadership.

Brooke's relationships with the commanders in the field were no less important and no less illuminating. There was none of the doubt that there had been between Robertson and Haig as to which was calling the tune. Nor was there any of the distrust which the men in command felt for Henry Wilson. All the Commanders-in-Chief and the Army Group Commanders, Auchinleck, Alexander, Montgomery, Giffard, Leese, and Slim knew exactly where they stood with Brooke. He gave the directions but did not interfere with their conduct of the battle. They knew they could speak their mind to him but that they would have to bow to his judgment—and they trusted that judgment. Montgomery always knew what he wanted to do and he always thought he was right—he usually was—but Brooke was the one man whose judgment he never questioned. Brooke's relationship with the Supreme Allied Commanders, Eisenhower, Mountbatten and Maitland Wilson

---

[24] Bryant, *The Turn of the Tide* and *Triumph in the West* (Collins, 1957 and 1959).
[25] Said in a lecture to a small group of officers of which the author was one.

The Strategy of the Second World War

had to be rather less direct because of the political and interservice factors, but they all had for him the same respect and trust that the army commanders had. In the War Office there was the same authority and mutual trust. Brooke's relationship with Grigg, the Secretary of State for War, and the other members of the Army Council, particularly Nye, the Vice Chief and veritably an *alter ego*, was such that he was free to give his whole energies to his Chiefs of Staff duties and yet remain truly the professional head of the army.

Germany's attack on Russia in May 1941 had brought to Britain a powerful ally, but it was by no means certain in 1941, or even in the early part of the summer of 1942, that Hitler would not be able to dispose of Russia and turn against the West. But the entry of the United States meant that victory for the Allies was only a question of time—barring serious mistakes on their part, and notwithstanding the fact that at the same time their enemies were joined by the Japanese with all their strength and fanatical tenacity. The turning point came before the end of 1942: first with the British victory at El Alamein; then the Russian victory at Stalingrad; and later the American victory at Midway Island.

Immediately the Americans entered the war Churchill went over to Washington to meet Roosevelt so as to make a combined review of the future conduct of the war. One most important result of Arcadia, as this conference was called, was the confirmation of an earlier tentative agreement giving priority to the war against Germany over operations against Japan.[26] From the United States point of view this was a brave decision since it was the Japanese attack that had brought her into the war and because to the American people the Japanese seemed to be the enemy on their doorstep. A significant result of this decision on priorities was the American determination that no time should be wasted and that since Germany could not be defeated except by a landing in North-West Europe, every effort should be directed towards preparing that venture as soon as possible. The Americans hoped it might be possible in 1942 but if not they were certain it must be done in 1943. Churchill and Brooke both knew that sooner or later the British had to come to grips with the Germans in

[26] *Grand Strategy*, vol. III, pp. 669-672.

Germany but both, and particularly Brooke, knew that probably the only way they could still lose the war would be by a bloody repulse in a major cross-channel operation.[27] They were determined that no attempt should be made until they were ready with all the strength necessary for such a hazardous venture. There were three major requirements: firstly victory in the Battle of the Atlantic so that the necessary concentration of the United States forces could be safely achieved; secondly the provision of sufficient landing craft and devices to make possible a landing in the face of strong opposition; and thirdly a degree of air superiority which would enable them to cloak their intentions and to move across the Channel with the vast armada necessary. Brooke saw that this would take time and although he did not wish to undertake any subsidiary operations that would delay the invasion (which was finally named Overlord) he was quick to see that in 1942 and possibly 1943 the choice would be between doing nothing and undertaking an operation in Southern Europe. He also emphasized from the first the importance of opening up the Mediterranean; this would save the million tons of shipping which the use of the Cape route added to their requirements[28] while the occupation of North Italy and the implied threat to South-East Europe would affect the German ability to resist in France.

Churchill's views are less easy to assess than Brooke's. He had such an instinct for diversions and was so resolved to strike the Germans everywhere at once that Brooke was hard put to it to restrain him from dispersing the available resources on minor enterprises.[29] Nevertheless Churchill did realize that it was heresy to the Americans to suggest that it was not possible to land in France in 1943. He therefore kept reverting to the major project and was fond of saying that his operations in the Mediterranean were a springboard and not a sofa.[30] Churchill was always interested in taking action in the Eastern Mediterranean islands and in south-eastern Europe but there is no evidence that such operations were put forward as an alternative to Overlord. Despite the British understanding of

[27] Bryant, *op. cit.* vol. I,, pp. 354-355, 357-362, 528-535.
— *op. cit.* vol. II, pp. 54-55.
[28] *Ibid.*, vol. I, p. 530.
[29] *Ibid.*, p. 340.
[30] *Ibid.*, 529-530. Churchill, *op. cit.*, vol. IV, p. 583.

the importance of Overlord there is no doubt that their early doubts about its practicability in 1943 and the consequent involvement in operations in North Africa, Sicily, and Italy did make Marshall and others suspect wrongly that Britain's heart was not in the cross-channel operations and that they had some nefarious designs in the Balkans.[31]

Even before the entry of the United States into the war Churchill had tried to interest Roosevelt in possible operations in North Africa and the Mediterranean.[32] Marshall, the United States Chief of Army Staff, soon showed his determination that nothing should divert the Allies from the assault on Europe. The Germans at this time had thirty-three divisions and strong air forces in Western Europe and Brooke's shrewd questioning[33] and a most pertinent note from Churchill to Roosevelt in June 1942[34] convinced the Americans that there could be no invasion in 1942. It then became clear that if anything was to be done to clear up the situation in North Africa preparations should be made at once. In the Desert the battle of Gazala had been lost and Auchinleck had stabilized the position in front of El Alamein with the Axis within what appeared to be striking distance of Alexandria and the Nile. Roosevelt not only arranged strong material support for the British forces in Egypt but also concurred in Churchill's cherished plans for an operation with the immediate aim of seizing the whole of French Morocco, Algeria and Tunisia, if possible with the co-operation of the French authorities there. This was the one occasion on which Roosevelt overrode his military advisers.[35]

The Americans were given the leading role in the African venture (called Torch) because it was thought more likely that they would get French co-operation. Eisenhower, who had been Marshall's Chief of Plans and Operations, appears in the resulting operations for the first time as an Allied commander. The reluctance of the American Chiefs of Staff to penetrate deep into the Mediterranean, a reluctance which Eisenhower did not share,[36] meant that the easternmost landing was at

---

[31] *Command Decisions*, ed. Greenfield (Methuen, 1960), p. 188.
[32] Churchill, *op. cit.*, vol. III, p. 576, and Butler, *op. cit.*, vol. III.
[33] Bryant, *op. cit.*, vol. I, p. 358.
[34] Churchill, *op. cit.*, vol. IV, p. 342-343.
[35] *Command Decisions*, p. 129.
[36] Bryant, *op. cit.*, vol. I, pp. 490, 494.

Algiers. The Germans thus had time to establish themselves strongly in Tunisia. However, Montgomery's victory at El Alamein had already been achieved and the pursuit begun before the Torch landings took place and the converging advance of the two armies soon removed all fears that the Germans would be able to retain a foothold. By May 1943 the whole of North Africa was in Allied hands and on 10 July began an attack on Sicily, the largest assault landing in the history of war.

While the battle for Sicily went on the Allies debated how best they could ensure the downfall of Mussolini and how to take advantage of his fall when it came about. The disadvantage of the Allied position was that if Hitler foresaw the defection of Italy he could exert a stranglehold on her whereas any attempt by the Allies to land further north than Salerno would be outside the range of land-based air cover. It was because of this distance from the heart of Italy that the British joint planners had tried at the Washington Conference in May 1943 to persuade Brooke to support an operation against Sardinia instead of Sicily. Brooke had been quite certain that the correct course was to open up the Mediterranean as quickly as possible. This was the purpose that had brought the Americans to North Africa. Brooke had gained American agreement to the Sicily operation with some difficulty at the Casablanca Conference in January 1943 and he knew that endless difficulties and suspicions would be raised if he now tried to switch the next operation to Sardinia.[37] If the Allies had determined on the conquest of Italy as their long term aim it might well have been better to go to Sardinia from Tunis, but the long term aim was the landing in France and clearly the Americans would not agree to any operations which seemed to detract from that aim. As it was we drifted into the operations in Italy and were forestalled by the Germans. Then for the remainder of the war the Italian Campaign was a bone of contention between Britain and the United States.

An important result of the absence of a common purpose in the Mediterranean was the shortage there of assault shipping and landing craft. Britain had invented the tank landing ship,

---

[37] Bryant, *op. cit.*, vol. I, pp. 543, 557-558.

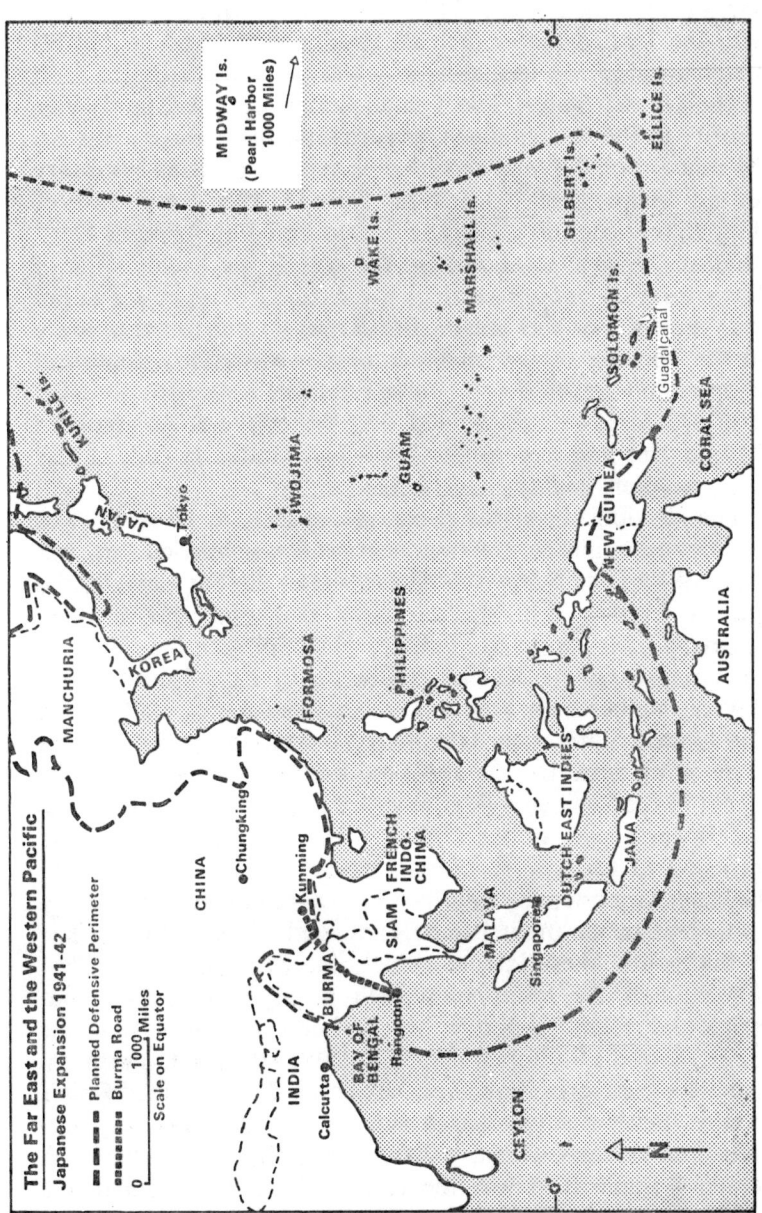

Map 9.

based on the idea of the old *River Clyde*,[38] but she had agreed with the United States that all landing craft, and indeed all merchant shipping too, should, while the war lasted, be built in the United States. The United States were willing to give priority to the war against Germany and so to make available the amount of shipping and landing craft considered necessary for an invasion of North-West Europe. But every stick of shipping outside this allotment was cornered by Admiral King, Chief of Naval Operations, for the war in the Pacific. One of the primary elements in strategy is the choice of an object within your means. For this reason Brooke was right to concentrate on the opening of the Mediterranean with willing American co-operation rather than to force on them a decision to delay Overlord to 1944 and to plan first the conquest of Southern Europe. But long before the Allies had landed in Italy it was clear that there could be no Overlord in 1943.

The relationship of the campaign in Italy to Overlord was to crop up again before that operation was carried out, but before discussing that we turn to events in the Far East. By the end of March 1942 the Japanese had gone far towards achieving their immediate military object of securing the great arc comprising the Kuriles, Gilbert and Ellis Islands, New Guinea, Netherlands East Indies, and the Andaman Islands. The United States view was that this situation could only be reversed by defeating the Japanese fleet, so the war had to be fought in the Pacific by their navy—and air forces were a large and integral part of the United States Navy. The war was entrusted to two commanders-in-chief: Admiral Nimitz with the fleet supported by Marines and a relatively small army contingent in the main Pacific area, and General MacArthur in the South West Pacific, which included Australia, New Guinea, and the Philippines, with a larger land force but substantial naval support. MacArthur's objective was the recapture of the Philippines and a possible advance to Formosa. Although these operations were subject to the Combined Chiefs of Staff they were regarded as solely a United States responsibility, and as was customary the commanders were given much greater latitude than was the case with British commanders under our own Chiefs of Staff. The operations therefore

[38] See p. 151.

The Strategy of the Second World War 205

concern the British conduct of the war only in so far as they affected priorities for the supply of amphibious shipping and landing craft, and the British were often left in the dark as to what was happening. The attitude of Admiral King has already been mentioned; he obeyed the instructions of the Combined Chiefs of Staff (of which he was a member) on the allotment of craft for Overlord but the Pacific war was his war and everything that was not specifically ordered away from him went to that war. Before leaving the Pacific Theatre it may be mentioned that the United States developed there a completely new technique of amphibious operations and the application of sea-borne air power. By the end of 1942, principally as a result of the Battles of Coral Sea and of Midway, they went far towards achieving the naval and air superiority which was the essential element in the destruction of Japanese power.

Although the principal American effort in the War against Japan was in the Pacific they did not divorce themselves entirely from the war in South East Asia, a war which by the loss of Malaya, Singapore, and Burma had brought the enemy to the gates of India and within striking distance of Ceylon. Japan had been engaged in China since 1937 and had almost half her army in China. The Americans with their eye on the eventual necessity of an attack on the Japanese homeland were anxious that China should continue to absorb this Japanese effort but feared that unless China received substantial aid she might seek a separate peace. The Allies could reach China only by the air route from Northern India over the Himalayas known as the Hump route. The Americans were therefore prepared to put substantial air forces into India and to co-operate with Britain in the defence of India. Moreover, the difficulties of the air route would be eased by the use of airfields in Northern Burma. The supply situation could be eased even further by improving rail communications from India into Assam and constructing a road and pipe line from railhead at Ledo to join up with the Burma road into China near Bhamo. The Americans were therefore willing to help the British in a campaign to recapture Northern Burma and would themselves provide the resources to build the road. They were not interested in going further south into Burma because they saw the Burma road and China as on the direct route to Tokyo while action towards Rangoon and Malaya would be a diversion.

The British view of China was different. They did not believe that China, having fought single-handed for five years, would contemplate a separate peace now that she had been joined by two powerful allies. On the other hand they did not foresee Chiang Kai Shek taking any immediate active part in the war and preferred to use their resources in a more direct blow against Japan.[39] The British view of operations in South-East Asia too was different from the American. They were naturally anxious to regain the lost British territories but neither Brooke nor Wavell, then Commander-in-Chief in India, wanted to get involved in a campaign in central Burma with all the difficulties of communications that would arise.[40] When resources became available the proper way to tackle the South-East Asia problem was by amphibious operations against Rangoon, Sumatra, and Malaya. By that means a real blow could be struck against the Japanese Army, for it was only in these areas and the Philippines that there were substantial land forces within reach. Churchill was already thinking in May 1943 of amphibious operations to seize one or more points on the crescent including the Andaman Islands, the Kra Isthmus, Sumatra, Java.[41] Brooke mentions that a few months later Churchill was also toying with the idea of operations against Akyab, Ramree Island, and Rangoon.[42] Brooke knew that such operations could only be done at the expense of the Mediterranean theatre and with some risk of delaying Overlord and that they should therefore be deferred until 1945.[43] Roosevelt and the American Chiefs of Staff pressed for the North Burma operations and wished also for an amphibious operation of some sort because Chiang Kai Shek believed that such was the most effective means of drawing off the weight of Japanese air effort against him.

Finally agreement or at least compromise was reached. An Allied command was set up in South East Asia. The selection

---

[39] For Brooke's view on Chiang Kai Shek see Bryant, *op. cit.*, vol. II, p. 79.
[40] Bryant, *op. cit.*, vol. I, p. 616.
[41] Churchill, *op. cit.*, vol. IV, p. 705.
[42] Bryant, *op. cit.*, vol. II, pp. 43-44.
[43] *Ibid.*, p. 46.

of Mountbatten as Supreme Allied Commander with Ceylon as his headquarters showed the bias towards amphibious operations. The disadvantage of the decision was that it emphasized the difference between the American and the British point of view and this was confirmed by the fact that Mountbatten received his orders and instructions from the British Chiefs of Staff. Thus while the control of the war against Japan and the allotment of resources was theoretically under the direction of the Combined Chiefs of Staff, there were in fact two separate efforts, one under the United States Joint Chiefs of Staff in the Pacific and the other under the British Chiefs of Staff in South East Asia. But because of the preponderant United States resources, the Pacific war was entirely run by the Joint Chiefs of Staff, or rather by King, with no British influence, while the British effort in South-East Asia depended to a large extent on American air forces and was considerably hampered by the meagre resources of assault shipping. Mountbatten's difficulties were much increased by the refusal of the United States to put those of their air forces that were primarily engaged in supplying China across the Hump under his command. The difficulty of having this division of command in a theatre which depended almost entirely on air transport to overcome the enormous administrative problems of overland operations in Burma would have been much more serious had it not been for Mountbatten's initiative and genius in getting people to work with him.

We know now what was not known in 1942 and 1943: that the Japanese did not intend to invade India. As it was the British desire to defend India and the American desire to secure North Burma worked towards the same end. The Japanese, unsettled by the ease with which the first Wingate Expedition had passed through their defences on the River Chindwin,[44] played into Slim's hands by seeking to seize a stronger defensive line on the Assam plateau. In this more open ground Slim inflicted a resounding defeat such as would have been almost impossible on the Chindwin position. Despite the geographical and administrative difficulties Slim went on through the centre of Burma to the Irrawaddy Valley and on to Rangoon. In doing so he inflicted the greatest defeat the

[44] See p. 278.

Japanese Army suffered during the war by virtually destroying the Burma Area Army. The details of that campaign are discussed in a later chapter but here it might be said that it was the counterpart of the Pacific war in that it showed the proper application of air power to land warfare. This was not a campaign in which armies and air forces fought in co-operation but a single integrated operation in which Slim and Baldwin[45] planned, as it were with a single mind, and used to a single aim, such resources as were at their disposal on land or in the air.

June 1944 was the month of Anglo-American victories. A secure footing was established in Normandy, Rome was captured, and victory in the Battle of Imphal was assured. In the same month the Russians started the offensive which was soon to reach the outskirts of Warsaw, an advance marred by the tragic consequences for Poland of the fighting within the city. The defeat of Germany was now only a question of time and the strategic problems facing the Allies from now onwards were of a different nature. In Western Europe there were two questions. One was the extent to which operations in Italy might still be useful or whether the effort could be better directed to Southern France. The other was the nature of the drive into the heart of Germany; whether it was to be a single strong thrust north of the Ruhr or an advance on a broad front to the Rhine with an encircling movement round both sides of the Ruhr. This second question is so much concerned with the tactical battle that its discussion is reserved for Chapter 12.

The relative importance of the battle in Italy revived all the controversy between British and American views of the war in Europe. The original conception had been that there should be a landing near Marseilles, to be known as Anvil, in conjunction with the landing in Normandy and at about the same time. This plan had been settled at the Teheran Conference in November 1943. It had the rather grudging agreement of Brooke, who thought that a continuing offensive capacity in Italy combined with the inherent threat of landings anywhere in the Mediterranean would be the best way of drawing German troops away from the vital area of Normandy.[46] As the date for

---
[45] Commander 3rd Tactical Air Force.
[46] Bryant, *op. cit.*, vol. II, pp. 91-176.

Overlord approached Anvil became even less desirable; the landing craft earmarked for it were required for the increase in the number of assault divisions for Overlord from three to five, and the situation in Italy, with Rome still uncaptured, was far different from that expected in January. Brooke became definitely opposed to Anvil and Churchill had come round to his view.[47] The Americans had been obliged, because of the landing craft question, to postpone Anvil but, despite the fact that it could not be carried out in concert with Overlord, they decided that the landing, now renamed Dragoon, must take place in July or August. This meant that Alexander had to withdraw seven divisions and prepare to mount Dragoon at the moment when he should have been able to concentrate all his forces and his activities on the exploitation of his victory in the battle for Rome. Churchill and Brooke decided it was better to accept this harmful decision than to cause a major breakdown in Anglo-American relations. The American determination to land in the south of France had two sound foundations: the advantage of opening up a Mediterranean port to increase the rate of build up of their forces in France, and the desire to get the Free French forces, which had played a truly magnificent part in the fighting in Italy, into France by the easiest route. By the time Dragoon took place the first requirement could be just as easily fulfilled by the use of the Atlantic ports in Brittany. Once the effect of the decision on the operations beyond Rome became inevitable Brooke no longer so strongly opposed Dragoon. Churchill, however, strongly supported by Alexander,[48] continued to press for its cancellation. Both were thinking of the possibility of a quick breakthrough into the Po Valley and an advance through the Ljubljana Gap to Vienna. No argument would be calculated to be more frightening to the Americans who imagined and distrusted awful British designs in the Balkans.[49] Brooke knew this and also knew that the advance towards Vienna was not a practical possibility except in the event of a crumbling of German resistance throughout Europe.[50] In fact although there was much talk of this advance through the Ljubljana Gap, the operation was

---

[47] Churchill, *op. cit.*, vol. IV. Appx D.
[48] Churchill, *op. cit.*, vol. IV, Appx D, pp. 656-662.
[49] *Ibid.*, vol. VI, p. 604 and *Command Decisions*, p. 294-295.
[50] Bryant, *op. cit.*, vol. II, pp. 224, 226, 256.

never planned in Alexander's Headquarters and it is not easy to see the situation in which it might have been possible. As Brooke so clearly saw, the Italian campaign was not an end in itself, but was a means of dispersing the enemy resources by engaging forces which could otherwise have been used in France or in Russia.

The postwar enmity between the Soviet and the Western Allies has tended to give a false idea of what we ought to have been doing to forestall Russia in Eastern Germany and South-East Europe. It is true that Churchill and Brooke both mistrusted Stalin and that the Russian attitude to the Free Polish Government and their action outside Warsaw gave great cause for mistrust. But it is as well to remember that the Russians were allies on whose success we placed great reliance in 1944. At the time of the Yalta Conference in February 1945, the Russians were within striking distance of Berlin while the British and American armies had not reached the Rhine. Two months later when Eisenhower had a small bridgehead over the Elbe, fifty miles from Berlin, the Russians were still on the Oder, the same distance away. The situation that faced Eisenhower was that the Allies had already agreed on the zones to be occupied by each nation and the Russians had been given as far west as the Elbe, although this decision was not to be allowed to restrict the operational moves of the armies. Montgomery had been pressing all along for a drive north of the Ruhr towards Berlin with the object of destroying the enemy in the North German plain. Brooke agreed with this strategic concept. Eisenhower, advised by Bradley, estimated that the advance to Berlin would cost 100,000 casualties.[51] He did not consider that in the light of the political settlement anything would be achieved by the loss of these American lives. Moreover, he was obsessed by the possibility of Hitler withdrawing with a hard core of his armies to fight in a last redoubt at Berchtesgaden and he moved part of his army to prevent this. The first that Churchill heard of Eisenhower's decision to halt at the Elbe was a copy of a message sent by Eisenhower direct to Stalin so informing him. Churchill was horrified and made an immediate request to the President and to Eisenhower for an effort to meet the Russians as far east as

[51] *Command Decisions*, p. 378.

possible. The President was in his last illness and died a few days later and Marshall could not be prevailed upon to interfere with Eisenhower's conduct of the campaign. One assurance Churchill did get from Eisenhower: Montgomery, deprived of the United States Ninth Army which had been under his command for the Rhineland battle and the crossing of the Rhine, would still be strong enough to seize the Baltic ports commanding the approaches to Denmark, and if he looked like getting into difficulties Eisenhower would lend support. By the end of April Montgomery had cleared Holland and Northern Germany and had captured Lübeck, thus holding the approaches to Denmark.

Eisenhower can be defended for halting on the Elbe. A soldier is at all times subject to political control and Eisenhower cannot be said to have failed to carry out his directive to ' undertake operations aimed at the heart of Germany and the destruction of her armed forces '.[52] His decision to halt Patton on the borders of Czechoslovakia was less reasonable. Eisenhower's Headquarters knew on 5 May that Czech partisans had risen in Prague and that German forces were attempting to deal with them. A Czech brigade was part of Patton's army and Patton could certainly have reached Prague before the Russians. Churchill tried to intervene on 7 May. This was after the armistice had been signed but as the Germans continued to attack the partisans there was no reason why Patton should not have been allowed to go on. The Russians entered Prague on 12 May and some eighteen days later gave permission for the Czech Brigade to enter the city.

The war with Germany ended on 7 May 1945. A few months before it had seemed that after the defeat of Germany the military weight of Britain must be moved to the war against Japan and that an arduous series of campaigns still lay before them. But Japan, defeated in the Pacific, in Burma, and in the Philippines, was on the verge of submission. The dropping of the two atomic bombs at Hiroshima and Nagasaki, whether or not they were necessary or justified, removed all doubt from the mind of the Japanese rulers. Japan surrendered on 15 August and the war was over.

A review of the strategy of the war from the fall of France

[52] *Command Decisions*, p. 377.

to the end reveals no major strategic mistake at any rate as far as the land operations were concerned. For much of this time the weight fell on British shoulders and the credit must go to Churchill and his military advisers, Dill and Brooke. Brooke showed himself as the outstanding architect of the war against Germany. His ability to stand the velocity of Churchill's tempestuous approach to strategy and to guide that great vision towards the attainable, and his tussles with Marshall as to the timing and direction of the great American effort in Europe, together made a contribution to victory such as has certainly not been made since Marlborough. Naturally he did not always get his way and, just as naturally, he was not always right. He was probably right in stressing the inexperience and lack of strategic insight of Eisenhower in the early stages—from the first he emphasized his personal qualities and outstanding ability to make allied armies work as one. But he hardly appreciated the speed with which Eisenhower learnt and how different a man he was in the final stages of the war. Again, hindsight can hardly justify Brooke's harsh criticism of the way the Americans worked the priorities between Europe and the Far East. It is true that the Americans were niggardly in the allotment of landing craft for the first priority, the war against Germany, and that this worked hard against Brooke's far-seeing attitude to the campaign in Italy, but the Americans did make magnificent use of their resources in the Pacific. If Brooke had had his way the war with Germany might have been over in 1944, but no one can tell what might have happened if the Japanese had been left in firm possession of their outer ring of defences until the war with Germany was over.

Chapter 11

TACTICAL ASPECTS OF THE WAR IN THE DESERT

It would hardly be profitable to take each battle of the Second World War in turn in order to discuss the tactical lessons. It is possible, however, to pick out a few battles and to see from them the extent to which the British had learnt from the Great War, the soundness of their tactical thought between the wars, and the reasons why their commanders failed or triumphed. The dividing line between strategy and tactics is never easy to define; the last chapter was confined to what is usually called 'Grand Strategy', and here the strategy of the battlefield is considered as one with the tactics.

It is worth considering why the conditions of stalemate which bedevilled three of the years of the Great War did not recur. The thoughtless answer is usually that it was because of the tank and the aircraft. This is very much a half truth. In battle the hidden anti-tank gun was always the master of the tank and aircraft never yet turned out determined troops from a well prepared position. Another answer is that British commanders were wiser because they were forewarned and were also more expert in their profession. That is certainly true, but war is a two sided contest and the Germans also had magnificent fighting commanders. They were no less well versed in cunning defensive devices and the use of entrenchments, mines, booby-traps, and obstacles than had been their fathers in the earlier war. The factor that changed was the size of armies. Industrial development had reached the pitch that required enormous man-power in war for the manufacture and maintenance of all aircraft, machines, and other equipment as well as for the complicated needs of civilian life. There were thus not sufficiently large armies to man in strength a defensive line across a continent. And so, even discounting the Western Desert, a model battlefield, most of the fighting took place under conditions where deception, surprise, and careful preparations could gain an early advantage from which the enemy could not easily recover. Nevertheless conditions not unlike those of 1918 often existed, for example, at El Alamein; throughout the Italian campaign; in the Normandy bridgehead; and in the bitter fighting in the Rhineland in February 1945. There the ultimate victory was achieved by means not very different from those of the Battle of Amiens. Throughout the second war it is clear that what are usually described as *Blitzkrieg* tactics did not succeed against a determined and well-commanded enemy. For success in battle it is nearly always necessary to mystify and mislead the enemy and to put its commander off his balance but, however well devised the plan, victory against a determined enemy requires in the last resort the will to overcome ' cost what it may '.[1]

The operations up to the fall of France have been discussed and it suffices here to mention the good showing of most of the senior British commanders in the field. Brooke, Paget, Alexander, Montgomery and Franklyn may be specially mentioned, and among the Brigadiers Dempsey and Horrocks showed the qualities which were to single them out for high command later.

In the winter of 1940/1941 operations in Egypt and Libya showed the Italians to have little stomach for the war and the British, Australian, and Indian regular forces to be in every way their superior. Even allowing for the advantage which this gave him O'Connor's campaign must be regarded as one of the greatest achievements in military history. With a force of only one armoured and one infantry division, totalling 34,000 men, he destroyed the whole of the 10th Italian Army, about 250,000 strong. Some 130,000 prisoners, 850 guns, and 400 tanks were taken at the cost of less than 2,000 casualties (500 killed, 1,373 wounded and 55 missing). The high standard of training of both divisions may be noted and the part that Hobart had played with 7th Armoured Division was a contributory factor, but it was the high pitch of efficiency with which O'Connor brought them to the battlefield and his handling of the battle that won the day. The careful preparations for the battle, the rehearsal of every detail, and the secrecy and deception which made surprise of the enemy complete, laid

[1] See p. 38.

Tactical Aspects of the War in the Desert 215

the foundations of success. In the battle all the components of the force—armour, artillery, infantry, and air forces—worked to the same plan and the same end. The interplay of the armour and the rest of the force is specially significant. The armoured division was always used to take advantage of its mobility and yet was always fighting where it could affect the main battle. At the end of one battle it was in a position which went half way to achieving the next object; O'Connor was always one step ahead. Finally, when the enemy could no longer offer organized resistance, an armoured column was sent in a daring dash across difficult desert country to hold the defile at Beda Fomm and so enable the rest of the force to complete the victory.

Throughout the operations the relationship between Wavell, the Commander-in-Chief, and O'Connor, the officer fighting the battle, was perfect. Wavell was always at hand and in control of the strategic situation and administrative support but never interfered with the conduct of the battle. Even the withdrawal of 4th Indian Division after Sidi Barrani to go to Eritrea, and its replacement by 6th Australian Division, which was a complete surprise to O'Connor, did not upset the harmony. Whether Wavell's decision was a wise one is a difficult point to judge. As a result of the change O'Connor's subsequent operations were delayed by about three weeks. Wavell's justification was that he took advantage of the availability of shipping to get a complete division to the Sudan in time for operations that winter, operations which were completely successful.

The memory of O'Connor's victory has been somewhat marred by the subsequent loss of so much of the territory he captured, in operations in which he was not in command, and his unlucky capture when sent forward by Wavell to restore the situation. Nevertheless, a study of these operations will show that the extent of the defeat was almost entirely territorial. The British force involved, although known as 2nd Armoured Division, was little more than a poorly equipped armoured brigade. Wavell must take the responsibility for a wrong appreciation of the situation since he over-estimated by more than a month the time Rommel would require before moving to the attack. Wavell's estimate was exactly in accord with the orders Rommel received from his superiors but Rommel was an opportunist of the first order. Wavell's handling of the deteriorating situation, his holding of Tobruk, and his steps to defend the Egyptian frontier were always competent and in some respects masterly. Rommel himself, referring to the British generals who opposed him, said 'the only one who showed a touch of genius was Wavell'.[2]

The failures in the desert in 1941 and the failures in Greece and Crete can be put down largely to shortage and inferiority of equipment and to inferiority in the air, with its serious repercussions on the sea situation. The subsequent British failures, the early disappointments in Auchinleck's 1941 offensive (Crusader), the second loss of Benghazi, and the severe defeat at Gazala demand another explanation. Throughout this phase the British had superiority in numbers both of men and tanks. No subject has caused more controversy than the relative performance and vulnerability of the opposing tanks and there is no doubt that the British came out of Crusader and the Battle of Gazala thinking that they had been let down by inferiority of equipment. The earlier models of cruiser tanks were mechanically less reliable than the German, but by Gazala much of this had been put right. Another weakness was the short range of the American Stuart or Honey tank with its fuel capacity of only forty miles. Apart from these disadvantages which might be balanced by superior numbers there is no reason to support the belief in inferiority of tanks. In armament and armour there was little to choose between British and German tanks in armoured brigades while the British infantry tank was much more heavily armoured than anything the Germans then had.[3] It may be unfair to take one incident but perhaps the surest guide to the fact that it was British tactics and not their equipment that was at fault is the fighting at Bir Gubi in 1941. There 22nd Armoured Brigade engaged Ariete Division, equipped only with Italian tanks, and was severely mauled.[4] For the Gazala battle British strength was much increased by the inclusion of Grant tanks, which amounted to about a third of their complement. The Grant had

[2] *Rommel Papers*, p. 520.
[3] Carver, *Tobruk* (Batsford, 1964), pp. 38, 39, 256-258.
[4] *Ibid.*, pp. 51-54.

faults in design which were ironed out for the production of the Sherman, but it was superior in gun and armour to anything the Germans had, even though they also had received tanks of an improved design. The appearance of the Grant was a nasty shock to Rommel.[5]

Although the capabilities of British and German tanks were not dissimilar, the Germans had much superior anti-tank artillery. Until Gazala the British had only the two pounder, the same gun as the tanks used. The Germans had the long barrelled 50 millimetre gun, much more powerful than the short 50 millimetre gun in their tanks and than the two pounder. Rommel also had a few 88 millimetre dual purpose (anti-aircraft and anti-tank) guns. The thirteen Rommel had had in Wavell's last unsuccessful offensive had shown how effective it could be. In the Crusader battle he had thirty-five, of which only twelve were with the Afrika Korps, the remainder being forward with the Italians on the frontier where there was no tank battle. At Gazala Rommel had forty-eight.[6] This gun, which could knock out any British tank at 2,000 yards, played a part in the battles far out of proportion to the small numbers available. That it was allowed to do so was a reflection on British tactics, since the gun was cumbersome and vulnerable and required considerable digging and labour even to hide it partially.

The British had a gun of about equal capacity to the 88 millimetre, the 3·7 inch anti-aircraft gun, but it was never extensively used in the anti-tank role. This was not because the Eighth Army had never asked for it but because at heart British gunners disapproved of a dual purpose weapon and because G.H.Q. and the Naval and Air Commanders-in-Chief were unwilling to release any from their essential anti-aircraft tasks in ports, airfields, and base installations. When at last G.H.Q. did relent and send some forward for anti-tank use they were sadly misused, in fact thoroughly despised by the gunners as being impossible to hide because of their height and the pillar of sand which followed the discharge of every round and so invariably drew enemy artillery fire.[7] As the

[5] *Ibid.*, p. 166, and *Rommel Papers*, p. 206.
[6] Carver, *op. cit.*, pp. 28, 39, 167.
[7] *Ibid.*, p. 211.

problem of siting the 3·7 was almost identical to that of the 88, this must serve to illustrate the inflexible approach of the British to tactics.

Effective as was the use the enemy made of his 88s, the framework of the German anti-tank defence was the long 50 millimetre, a gun which was not quite the equal of the British six pounder. It was issued to all types of German unit on a more lavish scale than was the British two pounder, and was always effectively and aggressively used in both attack and defence. As always, the Germans had a simple tactical drill which could be modified to suit the particular circumstances of a battle. They liked to choose the evening to close to contact because the defenders were much hindered by the low sun in their eyes. In the van would be light tanks and infantry in half-track carriers, while on the flanks would be anti-tank guns, and probably some 88s in the rear. When the advance had drawn British tank fire, often prematurely, the guns then destroyed the tanks thus located. Similarly in defence German anti-tank guns engaged British tanks before they got close enough to be dangerous. The broad principles on which Rommel worked was that anti-tank guns should destroy tanks, tanks should destroy infantry and unarmoured vehicles, and artillery should destroy anti-tank guns. This may be something of an over-simplification but certain it is that the Germans always used all their arms and weapons to a single cohesive plan. The simple but skilful minor tactics used by the Germans might not have been so successful if they had not been backed by a superb command technique, perfected in three major campaigns in Europe and a number of lesser campaigns and invasions. Not only was Rommel a tactical master but so were many of his subordinates, particularly Cruewell, commander of the Afrika Korps. Rommel was impetuous and would get involved in the forward battle so that he was away from his headquarters for hours or even days on end; his chief of operations, Westphal, was always ready, responsible and capable, to act in his place and give the command decisions. This professional technique went right through the Afrika Korps and was one of the main reasons why so often the British, groping for the situation, faced piecemeal the concentrated onslaught of the whole of Rommel's force.

The British therefore faced a tough proposition; the defeat of a highly trained professional army commanded by a master of the tactical art. Not all the advantages were on the German side, the British had much more experience of desert conditions, and individually the British and Indian soldiers—and indeed the Italian—proved much more adaptable and at home in the desert than did the German. Yet when it came to the fluid battle the British always seemed, in this phase, to be worsted. The fault which neither under O'Connor before nor Montgomery after ever existed, was that the armoured battle was considered as a thing apart. The task of the armour was assumed to be to seek out the enemy armour and destroy it and to leave the rest of the army to its own battle. We seem to have lost for the moment the idea that the battle was one and indivisible and necessarily subject to the design of one commander. This was perhaps a legacy of the pre-war controversy between armoured exponents and their more conservative brothers. So disgusted were the tank experts that they were hardened in their opinion that armour alone counted on the battlefield. Like a fleet at sea the armoured force would seek out the enemy to destroy it; the armoured force in being would remain supreme and ready to destroy the enemy headquarters and create havoc in his rear areas. The cavalry took some time to accept the idea of tanks instead of horses, but when they did they enthusiastically grafted on their own ideas of cavalry tactics. The supreme act of battle was again to be the cavalry charge, but this time with tanks. The most serious result of this separation of the battle between the armour and the rest was the dispersion of the artillery effort. Instead of Britain's well-developed practice of deploying the artillery so that it could be concentrated where it was most needed—for the support of the essential sector in attack or the vital area in defence—it was dispersed in columns or defensive boxes so that it was never strong enough anywhere. The course of the Crusader battle, although it finally resulted in success shows how an excellent plan and a brilliant preliminary advance to battle were marred in the execution by this underlying misunderstanding of the dependence of one arm upon another.

For Crusader the newly constituted Eighth Army, commanded by Cunningham who had the winter before conducted a successful advance into Abyssinia from Kenya, set out to destroy Rommel's army. The two sides faced each other across the Egyptian frontier, but Rommel used his Italian divisions for his defence while all his efforts were concentrated on the capture of Tobruk. For this purpose he had the Afrika Korps (two Panzer and one Light division) and one Italian armoured division. Cunningham had organized his army into XIII Corps, under Godwin-Austen, facing the frontier defences and XXX Corps, under Norrie, a primarily armoured force ready to strike round the southern flank. XIII Corps consisted of two infantry divisions and a brigade of infantry tanks. XXX Corps had three armoured brigades. These three brigades were all under 7th Armoured Division commanded by Gott, although strictly the division was made up of 7th and 22nd Armoured Brigade and the Support Group, while 4th Armoured Brigade Group was independent. XXX Corps also included 1st South African Division, which had been selected partly because its anti-tank guns (converted eighteen pounders) had a superior performance to the two pounder. The selection of this division, which had served under Cunningham in Abyssinia, was unfortunate because of its lack of desert experience and training. Auchinleck wanted to give this division more time in training but political pressure to begin Crusader was too strong to allow more than three days' delay. In addition to his two corps Cunningham had under his command the Tobruk garrison consisting of one division and one infantry tank brigade. Here it may be mentioned that a British armoured brigade had rather more tanks than a panzer division.

Cunningham's plan was devised to seek out the enemy and bring him to battle on ground not of the enemy's own choosing in order to destroy his armour. When that had been done the Tobruk garrison would break out and XIII Corps would advance so that between them they would roll up the remainder of the enemy. When Cunningham discussed his outline plan with his subordinate commanders, Norrie wanted to fulfil the first aim by advancing with the whole of his corps to Sidi Rezegh. He thought that Rommel must react to this threat to link up with Tobruk by moving to meet him. Norrie's plan met with objections from XIII Corps. Freyberg, whose national responsibility for the security of the New Zealand

Tactical Aspects of the War in the Desert

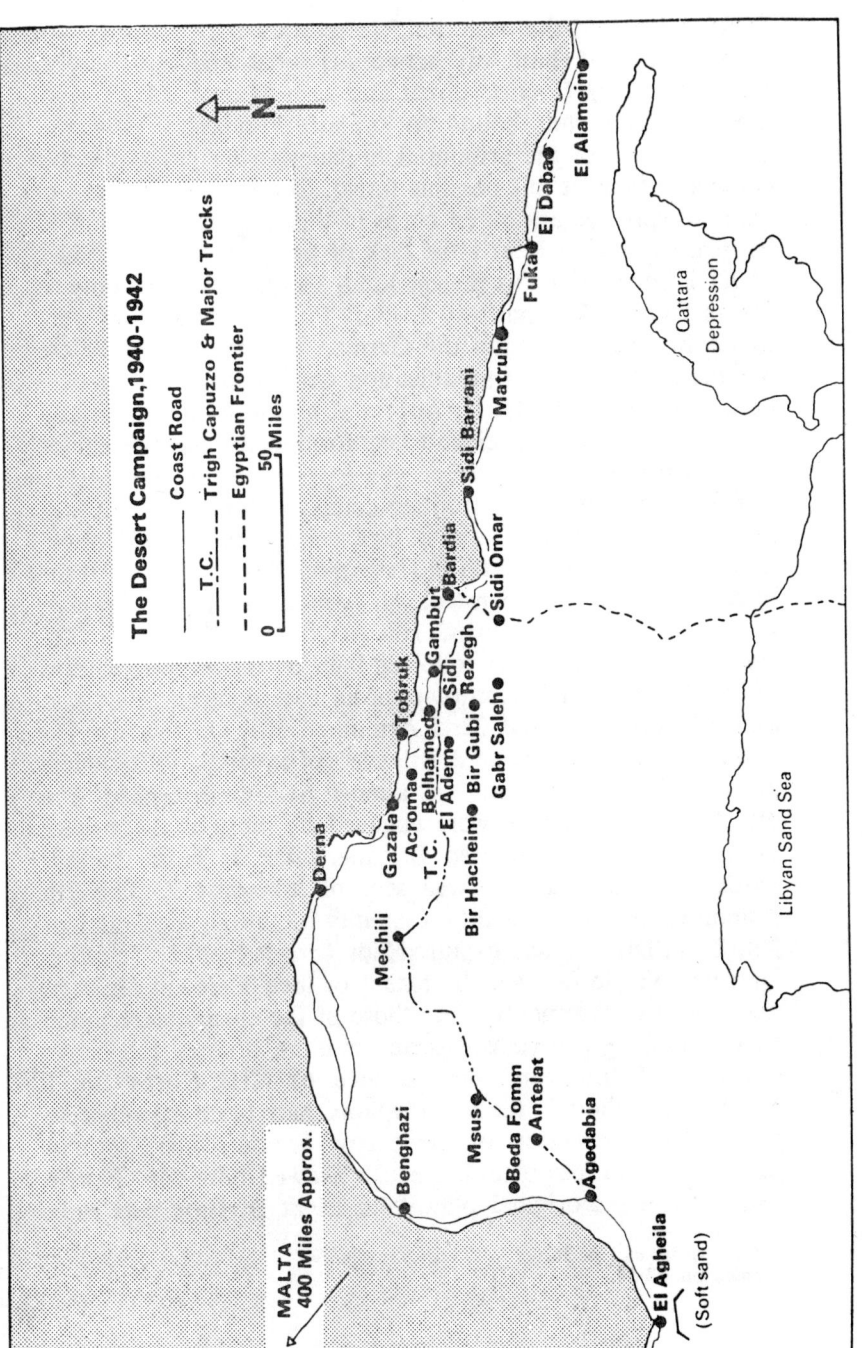

Map 10.

Division inspired the objection,[8] feared that the panzer group, then at Gambut, half way between Tobruk and the frontier, might disregard Norrie's threat and turn against XIII Corps. Godwin-Austen felt bound to support Freyberg's view and asked for one of the armoured brigades to be put under his command to meet this possible threat. Cunningham would not agree to split the armoured corps in this way but he modified his plan by giving Norrie the task of protecting the left flank of XIII Corps. For this reason XXX Corps was to go only as far as Gabr Saleh, halfway to Sidi Rezegh; there they were to see in which direction the German Panzers moved before seeking battle. By this modification Cunningham had surrendered the initiative and, far from seeking the battle on ground of his own choosing, decided to conform to the enemy's choice of battleground.

The move round the frontier defences to Gabr Saleh was faultlessly carried out. It was in fact such a surprise to Rommel that he failed to react at all. Thereupon Norrie sent his three brigades off widely separated; 4th Armoured Brigade keeping in touch with XIII Corps; 7th Armoured Brigade to Sidi Rezegh, where they were followed later by the Support Group of 7th Armoured Division (two motor battalions and field and anti-tank artillery); and 22nd Armoured Brigade towards Bir Gubi. The 22nd Armoured Brigade got involved with Ariete Armoured Division and lost almost half their tanks.[9] The 4th Armoured Brigade were engaged by most of the panzer group and 22nd Armoured Brigade, after its unsatisfactory battle with the Italians, was sent off to assist it. The 7th Armoured Brigade had little difficulty in getting the better of 90th Light Division and captured Sidi Rezegh. Neither Cunningham nor Norrie realized the extent of their losses in tanks in these early engagements while those of the enemy were much exaggerated by everyone. Norrie therefore ordered the break out from Tobruk and the advance of XIII Corps for 1 November although these operations had been intended to follow the destruction of the enemy armour. The German reaction to the loss of Sidi Rezegh showed how wise Norrie had been in his original plan. 90th Light Division was soon

---

[8] Carver, *op. cit.*, p. 36.
[9] *Ibid.*, pp. 53-54.

Tactical Aspects of the War in the Desert 223

joined by the two panzer divisions which had given the two British brigades the slip. In a confused series of actions on 22nd and 23rd November 7th Armoured Brigade was destroyed and the Support Group, fighting magnificently, was overwhelmed, while Rommel held off the remainder of XXX Corps with his anti-tank screen. In almost a chance encounter the rear echelons of 4th Armoured Brigade and 1st South African Division were overrun. Then the whole Afrika Korps turned on 22nd Armoured Brigade and one of the two brigades of 1st South African Division. This battle seems to have been commanded by Cruewell and the whereabouts of Rommel are obscure. Both British formations suffered grievously but they also inflicted great loss on the enemy and destroyed some seventy-two tanks leaving Cruewell with only ninety out of the 249 tanks which he had at the beginning of Crusader. At the end of the day on Sunday 23rd the full extent of his losses came home to Cunningham and he came to the conclusion that he had lost the battle. Rommel was even more certain that he had won. Cunningham decided to break off the battle and actually issued the orders for withdrawal behind the frontier in order to reorganize for a return to the offensive later. Galloway, his chief of staff, and the two corps commanders were certain that such a withdrawal could only lead to destruction and that the Germans, although they had undoubtedly had the best of the armoured exchange, had suffered losses too heavy to enable them to clinch the victory. Both Tobruk garrison and XIII Corps were going well and Godwin-Austen, with both his corps and the best part of his infantry tank brigade still intact, strongly opposed the idea of breaking off the battle. Galloway rang up Whiteley, Auchinleck's chief of operations staff, to insist that the situation required Auchinleck's presence forward before anything irrevocable was done.[10] Auchinleck flew forward at once, quickly appraised himself of the situation and courageously decided that the offensive must be continued. He ordered Cunningham to direct his operations to the recapture of Sidi Rezegh and linking up with the Tobruk garrison.

[10] *Ibid.*, p. 87 and *Operation Victory de Guingand* (Hodder and Stoughton, 1947), p. 98. Letter from Field Marshal Sir Claude Auchinleck to the author, 23rd October, 1965.

At about the same time that Auchinleck made his decision Rommel, seeing himself as the victor in the armoured battle, moved with all the tanks he could collect to encircle the British forces fighting on the frontier. Both New Zealand Division and 4th (Indian) Division had made considerable progress in their operations. 4th (Indian) Division had captured the Italian defences at Sidi Omar and Rommel made no impression on them. The New Zealand Division were well beyond the frontier and were linking up with the Tobruk garrison. Rommel was forced, therefore, to give up his foray into Egypt, and to return to restore the situation around Tobruk. Auchinleck had thus by his intervention for the first time gained the initiative in the battle.

Two significant points arise from this phase of the battle. First that it was XIII Corps with no armoured forces other than their infantry tanks who withstood the onslaught of Rommel's armour on this the only occasion when he departed from his practice of fighting with a force of all arms. The second is the question of command. No principle is more sacred in British military tradition than that of loyalty. In battle, initiative and willingness to accept responsibility are essential in staff officers and subordinate commanders but all is done under the single will and to the tune of the commander. In Crusader Galloway and Godwin-Austen were in a most difficult position in that they saw that victory still lay within Eighth Army's grasp while their commander was determined on action which would almost certainly lead to disaster. Great credit is due to Galloway for resolving this question by making clear to Cunningham his opinion and ensuring that Auchinleck came forward before it was too late. Before Auchinleck flew back to Cairo he had a long talk alone with Galloway. Auchinleck made it clear how much he trusted Galloway and relied on him to see that the battle was pressed to the uttermost.

No episode in the campaign illustrates more fully the true nature of battle. How difficult it is to define success and failure and how narrow is the line between them. How much less important are the movements of this and that division than what is going on in the minds of the commanders—of what each thinks the other is thinking—and in the minds of their men too. Crusader was an action in which the situation

Tactical Aspects of the War in the Desert

changed from moment to moment and in which the two armies faced each other in all the uncertainties of war. On the one side the principals were: Cunningham with his doubts when the extent of his losses became known; Galloway with his fearless understanding of the realities of the battle and all that was at stake, and his willingness to take responsibility; Auchinleck with his steel will and his inflexible determination; and below them an army resolute and ready to obey but, it must be confessed, an army of magnificent amateurs. On the German side: Rommel with his impetuous genius and his ability to seize the faintest opportunity; Cruewell with his real grasp of armoured tactics and his command machinery which could retain control in the most confused battle; Westphal, whose ability to act as the complement of Rommel has already been noted; and below them an army experienced in war and trained to a hair's breadth in all the technique of battle.

By his dash to the frontier Rommel had left XXX Corps free to reorganize from the ample reserves which excellent Q arrangements made available. The 7th Armoured Brigade appears no more but Norrie now had 4th and 22nd Armoured Brigades at reasonable strength again. Rommel was by no means finished and was able again to intervene between Tobruk garrison and New Zealand Division. But hard fighting here and the appearance of Norrie's corps on his southern flank persuaded Rommel that the battle was lost. He decided, despite the protest of his Italian superiors, to withdraw right back to El Agheila. In the meantime Auchinleck had relieved Cunningham of command and appointed in his place Ritchie, who was then serving as his own Deputy Chief of Staff. This was an unfortunate choice, not because of Ritchie's abilities but because of the relationship between the two men. Ritchie had seen much fighting throughout the Great War and as a junior officer had won the D.S.O. and M.C. in Palestine in 1918. In 1940 he had been Brooke's chief of staff in the fighting from the Dyle to Dunkirk and Brooke had absolute faith in him. His quality may be judged by the fact that after his removal from command of the Eighth Army in 1942 he assumed command of a division with all the enthusiasm of a newly promoted brigadier and was later a corps commander under Montgomery throughout the whole of the campaign in North West Europe.

The difficulty now was that Ritchie was appointed as Auchinleck's deputy to fight the battle according to his ideas. This is an impossible position because a commander must command in his own right not as a deputy. Auchinleck presumably did not want to interfere with the command structure by promoting one of the corps commanders and thus taking him from his immediate task and there was obviously no time to get someone out from home. In these circumstances Auchinleck ought to have carried on exercising personal command. If he had had to answer a call for a short visit to Cairo he had Galloway in control to carry out his intentions, but such absence would hardly be necessary. Arthur Smith his Chief of Staff in Cairo, and Wavell's before, was quite capable of running the affairs of Middle East while Auchinleck completed the battle. In the necessary pause at El Agheila there would be time enough to appoint a new commander. The situation after Ritchie's appointment is summed up in the words of the Official History:

> Thereafter Auchinleck spent days at a time close to Army Headquarters not interfering with Ritchie but ensuring that his own advice and support were available if they were wanted. All important operational matters were discussed between the two. It may be wondered whether in these circumstances there could have been much distinction between advice and orders.[11]

Originally Auchinleck had meant Ritchie's appointment to be only temporary but after arrival at El Agheila he decided to keep him in command. This despite the fact that the pursuit of Rommel had not been particularly well handled.

There was one fleeting moment when the Eighth Army might have smashed the Panzer Army once and for all. Rommel had brought forward every gun from Benghazi and packed everything forward for his last unsuccessful attempt to prevent the relief of Tobruk. All the information at Eighth Army Headquarters indicated that the Germans were in a bad way and could not deal with fresh troops getting in behind them. Ritchie still had 22nd Guards Brigade Group which, fresh and intact with its own armoured cars, tanks, and

[11] Playfair, *op. cit.*, vol. III, p. 97.

artillery, had not been committed to the battle. On 29 November Ritchie had warned Marriott its commander to prepare for a move across the desert to Antelat to cut off the German retreat. Miller, Ritchie's chief Q staff officer, was confident that this move was administratively possible and that our supply position was much better than the Germans'. Coningham, commanding the Desert Air Force, was enthusiastically in favour of the plan and had ready five or six squadrons for direct air support all the way. There was a nasty interlude from 1st to 5th December when Rommel had again separated XIII Corps from the Tobruk Garrison and 22 Guards Brigade stood by to support 4th (Indian) Division at Bir Gubi. However, Rommel saw that all was lost if he stayed and so on the night of the 6th he began to pull out. Galloway urged Ritchie that this was the moment to launch the brigade group. By 9th December it was clear that Rommel had gone and by the 11th he was at Gazala. At this stage 4th Armoured Brigade, refitted to a tank strength of nearly three times what the Germans had left, was back in the battle and Ritchie shelved his bold plan for the move to Antelat and substituted an attempt by XIII Corps to bring Rommel to battle at Gazala. Ritchie still kept 22 Guards Brigade Group in hand for the dash to Antelat but he did not use it until the opportunity of its achieving anything had passed.

This incident throws a light on the meaning of command in battle. By training and thought it is possible to learn to discern the opportunity but to act requires the flair and the courage to take the decision with all the risks that are involved. It is a moment of decision and of responsibility which the commander cannot share. So often the handling of Eighth Army can be criticized because armoured brigades were used in isolation where they achieved nothing; here was an occasion where a brigade group of all arms, strong in relation to what Rommel had left, might have played a decisive part in the battle. As it was 22nd Guards Brigade did not move off until 20 December; by then it was too late for anything but a direct pursuit. On Christmas Day 22nd Guards Brigade, now under command XIII Corps came up against the enemy at Agedabia. The 22nd Armoured Brigade had also joined XIII Corps and was operating some thirty miles south of

Agedabia. Cruewell had by this time been reinforced by two squadrons of tanks, all that had survived of a convoy of reinforcements by sea to Benghazi. Cruewell was able to see once more the fatal habit of British armour working on its own, and, despite an inferiority of two to three in tanks, seized the opportunity for a quick counter-stroke. In the engagement 22nd Armoured Brigade lost half their tanks after which the Germans continued their withdrawal to El Agheila.[12]

The confirmation of Ritchie in command meant that the two corps commanders had been passed over by an officer much junior and much less experienced. They were both men of too generous a spirit to let this affect their loyalty but the manner of it did lead to a lack of confidence in the orders they received. When Auchinleck was in control of the battle everything went well, but when he was back in Cairo nobody knew whether the orders they received were second-hand orders made without knowledge of the immediate situation and passed on from Cairo or whether they truly emanated from Ritchie. There thus grew up a habit throughout the Eighth Army of treating an order as a subject for discussion. No one really knew who was fighting the battle and there was certainly never a single mind in control. Time and again this attitude of mind was the source of trouble and it was certainly the greatest single factor in the loss of the Battle of Gazala and of Tobruk.

An earlier example which illustrates this lack of confidence in the command system came at the end of January 1942 when Rommel returned to the attack. He had received reinforcements of fifty-four tanks and their crews but at that time his total tank force numbered only eighty-four. The Germans were able to count on local air superiority and their air forces had prevented the British from developing Benghazi as an advanced base. Eighth Army was not administratively in a position to fight a major battle forward of Benghazi and was not suitably disposed or equipped for wide observation and delaying action. For example, all the armoured cars were in Egypt refitting. 1st Armoured Division had 2nd Armoured Brigade north of Agedabia and its motor brigade was forward watching the

---

[12] Playfair, *op. cit.*, vol. III, pp. 81-92; Liddell Hart, *The Tanks*, vol. II (Cassell, 1959), pp. 139-143; Marriott, *Military Memories* (privately published, 1960); Letters to the author from Lieut.-General Sir Alexander Galloway and Major-General C. H. Miller, Feb. 1968.

exits from El Agheila at the wide end of the funnel. 4th (Indian) Division was at Benghazi and the whole was under Godwin-Austen's XIII Corps. Rommel came forward tentatively and things went well for him; he had soon inflicted heavy casualties on 2nd Armoured Brigade and was in a position to by-pass or to strike at Benghazi. Godwin-Austen realized the difficulty of fixing Rommel in order to strike him and wanted to retire so that he could give battle at about Mechili. Ritchie, pressed by Auchinleck to retain the Djebel bulge and the airfields at Derna—both were important in the fight to defend Malta—would not agree and sent contrary orders direct to the divisions. The result was that Rommel was not brought to battle and the position was not stabilized until the British were back at Gazala, unpleasantly close to Tobruk as subsequent events were to show. Godwin-Austen felt that Ritchie had shown no confidence in him and asked to be relieved of his command. He was replaced by Gott who had fought with 7th Armoured Division under O'Connor, had commanded it in Crusader, and who, because of his courage, skill and experience was almost a legendary figure in the desert.

After this reverse Eighth Army set about making themselves secure behind the Gazala line and preparing once more for the offensive that was intended to destroy Rommel's army and drive the Axis out of Africa. Churchill, with the new weight on his shoulders of the Japanese successes in Singapore and Burma and with a realization of the danger in which Malta lay, pressed Auchinleck to go immediately to the offensive. Auchinleck was determined that he would not do so until he could be sure of a three to two superiority in armour and he doubted if this would be before the beginning of June. Despite the importance with which Auchinleck regarded the offensive he also realized that Rommel might strike first. The navy and the air force in the Middle East had been somewhat reduced in order to reinforce the Far East, and the three commanders-in-chief at Cairo had therefore made up their mind that if Rommel made the Gazala position untenable the Eighth Army would go right back to the Egyptian frontier because the drain of attempting the isolated retention of Tobruk could not again be afforded.

The chief danger from Rommel was that he might hook

round the southern flank and strike at Tobruk and Belhammed, a quickly growing administrative area on the safety of which the British offensive preparations depended. In order to prevent this our thirty-five mile long Gazala line was extended fifteen miles further southward to Bir Hacheim. Considerable work was done to strengthen the defences and much of the material used for this was obtained by dismantling the Tobruk defences. No more troops were available for the defence and so considerable gaps in the position had to be accepted. XIII Corps astride the coast road had 1st South African and 50th Division (less one brigade), supported by two brigades of infantry tanks, in a fairly compact position astride the coast road. There was then fifteen miles with only the third brigade of 50th Division and this was sited to bar the central route to Rommel. Then again there was a gap of fifteen miles to Bir Hacheim, held by the Free French Brigade Group, under XXX Corps. It was thought unlikely that Rommel would be able to break through along the coast road; Auchinleck thought it most likely that he would strike through the minefields in the centre and make along the route known as the Trig Capuzzo for El Adem and Tobruk; Ritchie was more inclined to think Rommel would come round the southern flank, and Norrie foresaw a stroke in the coastal sector combined with a simultaneous southern thrust. Auchinleck thought Ritchie should site all the armour astride the Trig Capuzzo and there await Rommel's attack. Ritchie did not like this plan because it would bring the armoured battle too near to Tobruk, the vital administrative centre for the future offensive. In the light of these conflicting views something of a compromise was reached. 1st Armoured Division (2nd and 22nd Brigades) was put astride the Trig Capuzzo and 7th Armoured Division (4th Brigade) further south, about twelve miles north-east of Bir Hacheim. Plans were made for 22nd Armoured Brigade to move south should Rommel come that way and it was hoped that wherever he came the whole weight of the British armour (Auchinleck had about his three to two superiority) would be brought against him. The remainder of Ritchie's army was placed in self-contained brigade group positions, known as boxes, well wired, mined, and dug in at such places as ' Knightsbridge ', Acroma, and El Adem where it was thought they would bar Rommel's progress and form hinges on which our armour could manoeuvre. Ritchie had 2nd South African Division in reserve at Tobruk.

Ritchie has been much criticized for his wide dispersion in his forward positions and because much of his wire and his minefield were not covered by fire. This is an unjust criticism; the extension to Bir Hacheim was forced on Ritchie by the build-up of the Tobruk maintenance area necessary for the offensive. It was better to have the wire and the mines where in the event they posed a considerable problem to Rommel, than to have unmined and unwired gaps. The weakness in the dispositions was in the rear areas where, instead of a compact position covered by the concentration of his artillery where he could fight the battle for the decisive ground, Ritchie had dispersed his force in 'boxes' which Rommel could either by-pass or destroy in detail. But the real weakness was not in the dispositions but in the way the battle was fought. Let Rommel himself pass judgment on the defences. Of them he said: 'Had there been German troops holding these British positions, they could hardly have been taken.'[13]

On 26 May Rommel made a feint towards the coast sector and then with the Afrika Korps and the Italian XX Corps (Ariete Armoured and Trieste Motorized Divisions) moved by night round the southern flank at Bir Hakeim. The plan was for the two panzer divisions to strike for Acroma defeating the British armour on the way. The 90th Light Division was to make a demonstration towards El Adem while XX Corps protected the left flank and Rommel's line of supply. This accomplished Panzer Group would continue to the sea and turning west would roll up the British forward troops while 90th Light held off reserves moving up from Tobruk. Rommel had underestimated the depth and extent of the British obstacles and the strength with which Bir Hakeim was held. Moreover the whole of his move was observed and reported by the South African armoured cars and light detachments from the motor brigade of 7th Armoured Division. Rommel got well behind his time-table and had little opportunity for surprising his opponents. Yet, by some failure in communication which has never been explained (it was not

[13] *Rommel Papers*, p. 457.

a signals failure), 7th Armoured Division was surprised. Its headquarters was overrun and 4th Armoured Brigade brought to battle at a disadvantage, suffering unnecessary losses. Then 22nd Armoured Brigade was caught by surprise preparing to move south and lost thirty tanks. Despite all this, by nightfall on the 28th the Germans were in a difficult position. The first appearance of the Grant tanks had been an unpleasant shock, Rommel had lost one third of his tanks and both panzer divisions had run out of fuel. He had lost touch with 90th Light, which had made little impression on the defenders of El Adem. XX Corps had made no progress in finding a way through the minefields and Rommel with no supply line was in a very vulnerable position.

No co-ordinated attack was made on Rommel on the 29th; that night he was able personally to lead up some supply lorries that had come all the way round, to bring back 90th Light, and to make a new plan. He then determined to drive a wedge through the centre of Ritchie's forward position, that is astride the Trig Capuzzo; XX Corps attacking from the west side while 90th Light attacked from the east and Panzer Group held off the remainder of Eighth Army. The 150th Infantry Brigade, the isolated brigade of 50th Division, bore the whole brunt of the attack; they fought to the last round but by 1st June they were overrun. During these days no co-ordinated major attack had been made on Rommel's main position. He was able to remain in the area which came to be known as the Cauldron, there he had a secure supply line and was free to turn all his energies on the destruction of the Free French at Bir Hakeim. At this stage Ritchie made the first attempt at a co-ordinated counter-attack using infantry and tanks. This attack, known as Aberdeen, was made by part of 7th Armoured and part of 5th Indian Divisions and one brigade of infantry tanks. Despite the fact that there were two corps commanders available neither of them took any part in the battle. The two divisional commanders, Messervy (7th) and Briggs (5th), took turns in command according to whether infantry or tanks were in the leading role.

In the knowledge of the dust and confusion of battle in the desert and the difficulty of knowing the situation or even of knowing friend from foe, it is unfair to simplify the battle.

Tactical Aspects of the War in the Desert 233

Nevertheless in order to draw the lessons it is necessary to do so. The plan was complicated; the first phase was a night attack by 10th Infantry Brigade supported by 4th R. Tanks, involving an advance of four miles. Then 22nd Armoured Brigade was to pass through and advance some five miles into the heart of the Cauldron. This would allow 9th Infantry Brigade, also supported by 4th R. Tanks and by four field artillery regiments brought up into 10th Brigade position, to advance to take and hold the German held gap in our defences. 22nd Armoured Brigade would then wheel north in an attack in conjunction with a strike south from XIII Corps area by a brigade of infantry tanks. An officer then serving on Norrie's staff describes the plan thus:

> If all went according to plan, the whole area occupied by Afrika Korps and Ariete would in this way be quartered with troops; but the broad arrows of the planners' maps would in fact represent a series of inexperienced battalions, weakly armed with anti-tank weapons, driving or walking over many miles of open desert to objectives which were several miles from each other. If ever an operation resembled sticking one's arm into a wasp's nest, this did; and, in its final version, it certainly did not produce the armour on the minefield gap in rear of the enemy.[14]

The 10th Infantry Brigade captured its objective, but unhappily this was short of the enemy position so that the supporting fire did no more than warn the enemy of the attack. Therefore the advance of 22nd Armoured Brigade was met by the concentrated artillery and anti-tank fire of the whole Afrika Korps. The attack by the infantry tank brigade isolated from this battle, ran into an unknown minefield and all but twelve of its ninety tanks were destroyed by anti-tank fire. Rommel was then free to launch his counter attack; in the confused fighting which followed both divisional headquarters were overrun and 10th Infantry Brigade, part of 9th Infantry Brigade, two other infantry battalions, and the four regiments of artillery were isolated and virtually destroyed.

The failure of Aberdeen meant there was little hope of holding Bir Hakeim, and, indeed, little purpose in so doing.

[14] Carver, *op. cit.*, p. 195.

The garrison was therefore ordered to fight its way out. Ritchie still had superiority in tanks with two hundred and fifty cruiser and eighty infantry tanks against Rommel's total of two hundred and twenty-six of which almost a hundred were Italian or light tanks. 1st Armoured Division was north and 7th south of Knightsbridge, where 201st Guards Brigade held one of those defensive boxes designed to serve as a pivot for armoured manoeuvre. Nevertheless by a series of misadventures, as often as not because one of the divisional commanders felt he had a better plan than that he had been ordered to carry out, the three armoured brigades were destroyed in turn by the concentrated attack of Rommel's tanks. At the end of the battle of Knightsbridge the Eighth Army had only seventy tanks in all. There was no longer any doubt in anybody's mind that XIII Corps, which had taken virtually no part in the battle so far, must be withdrawn. The earlier decision not to try to hold Tobruk as an isolated garrison will be remembered; but neither Auchinleck nor Ritchie wanted to lose Tobruk with all that this meant for the hope of a future offensive. To Churchill, too, the loss was unthinkable. He sent a message to Auchinleck presuming that there was no intention of giving up Tobruk[15] but there was certainly no question of his having issued instructions or interfered in the conduct of the battle. Auchinleck decided that Tobruk must be held because he still believed Rommel could be held on the line Acroma-El Adem. Ritchie knew that it was too late for this but thought that Tobruk might be held temporarily while the frontier position was occupied and a riposte prepared. At the last moment, therefore, Klopper was ordered to hold the dismantled defences of Tobruk and to prepare to co-operate with XIII Corps by active operations ' as far from the perimeter as possible '.[16]

In addition to his 2nd South African Division of two brigades Klopper had a brigade of infantry tanks, 201st Guards Brigade and 11th Indian Infantry Brigade under his command. Two of the Brigadiers, including the tank brigade commander, had been in the first siege. The uncertainties and the last minute orders were hardly conducive to a repetition of the splendid defence of 1941 and this time Rommel, on the crest of the

[15] Churchill, op. cit., vol. IV, p. 372.
[16] Carver, op. cit., p. 233. Ch. 11 is a detailed account of the loss of Tobruk.

Tactical Aspects of the War in the Desert

wave of victory, lost no time in going straight for his objective. More than either of these factors, however, it was the lack of a single mind conducting the defence that lost the battle. The commander and the various brigadiers all saw the battle slightly differently and so the hard fighting of individual units was of no avail. After a vain attempt to fight their way out the bulk of the garrison capitulated on 21 June.

It had been intended that after the capture of Tobruk the Axis forces would turn to the assault of Malta. Rommel's success was, however, more overwhelming than even he had expected and he was not the man to miss the opportunity of pursuit and perhaps bouncing through to the Nile Delta. Ritchie decided not to fight on the frontier but to make a conclusive stand at Mersa Matruh and also to send back Norrie to organize a position in depth at El Alamein. Two days after the fall of Tobruk, Rommel set off in pursuit with only forty-four tanks. On the Matruh position Auchinleck decided to relieve Ritchie of command and to take over Eighth Army himself. Despite the fact that a new headquarters, X Corps, had come forward with the New Zealand Division there was little cohesion on the Matruh position. Auchinleck therefore decided to go back to El Alamein where there was a position with closed flanks, running forty miles from the sea to the Quattara Depression. Here on 1st July Rommel was finally brought to a halt more than three hundred and fifty miles from Tobruk and only fifty miles short of Alexandria. Eighth Army was much heartened by Auchinleck's firm personal command and strengthened by the addition of 9th Australian Division as well as the New Zealanders. In hard fighting throughout July Auchinleck achieved the ascendancy and Rommel no longer had a hope of getting to Alexandria without a pause for replenishment and reinforcement. Auchinleck went over to the offensive on 10 July but gained little advantage for the casualties he suffered.

Auchinleck deserves the greatest credit for so effectively stopping Rommel but it cannot be said that he ever looked like turning the tables on him. Eighth Army fought magnificently in its separate units but there was never the cohesion essential for victory; in particular the infantry and the armoured divisions did not altogether trust each other and the artillery effort was so dispersed as never to be truly effective anywhere. Some writers have referred to Auchinleck's July battles as the 'First Battle of Alamein', possibly to belittle Montgomery's subsequent victory. The official names for Auchinleck's two offensives, that is the names used in the official history and by the Battle Honours Committee, are the First and Second Battles of Ruweisat.

It was not to be expected that Auchinleck could continue in personal command of the Eighth Army as well as being Commander-in-Chief Middle East. Churchill wanted Gott to become army commander and insisted despite Brooke's doubts; but Gott was shot down in an aircraft and killed on his way to take over command. Gott was one of the outstanding fighting commanders of the war and the esteem in which Eighth Army held him has already been mentioned. Carver describes him:

> His clarity of mind, his rock-like imperturbability and common sense, his readiness always to propose a course of action when others faltered or were in doubt, all these qualities combined with a truly Christian character had made him the oracle to whom all, both high and low, turned for advice at all times, but especially in bad.[17]

He was certainly the man who had so often got Eighth Army out of its difficulties but it is doubtful if he was the man so to arrange matters that these awful emergencies would be less likely to occur. He had perhaps so read the lessons of the Great War as to eschew the concentrated battle, the persevering dog fight of the type Montgomery was to fight at Alamein. How else can we explain XIII Corps's aloofness from the Battle of Gazala; Gott and others toying with ideas of an advance to Tmimi and Derna while Rommel waited in the Cauldron to be destroyed?

The selection of Montgomery to command brought to the Eighth Army a fresh mind and gave them a commander whose outstanding characteristic was the ability to see a problem clearly without over-simplification; a commander, moreover, with the certainty of purpose and the strength to impose his will. Auchinleck had willingly accepted Montgomery as com-

[17] Carver, *El Alamein* (Batsford, 1962), p. 29.

Tactical Aspects of the War in the Desert 237

mander Eighth Army but it was thought that he himself should move on to be Commander-in-Chief in India and he was replaced by Alexander. In Alexander's command of Montgomery there was a return to the old relationship between commander-in-chief and commander in the field that had existed between Wavell and O'Connor; there was no longer any doubt who was commanding the battle.

Montgomery's first care was to ensure that he could defeat Rommel's attack if it came, as was probable, before Eighth Army was ready to go to the offensive. He quickly surveyed the battlefield and decided that in the Alam el Halfa ridge lay the key to the defensive battle. Whether Rommel came on down the coast or, as seemed more likely, pushed through the more weakly held southern sector, he would be unable to advance while the British held the Alam el Halfa ridge in strength. Montgomery therefore asked Alexander for one of the divisions in training in Egypt; 44th Division was sent up and sited to defend the ridge. The bulk of the armour was also placed in hull-down positions for a battle for the ridge and the artillery sited primarily to the same end. There had been some increase in the number of six pounder anti-tank guns available but Sherman tanks had not yet arrived for the armoured units; broadly speaking therefore the only increase Montgomery had over the forces available to Auchinleck was 44th Division. The air forces were again brought back into proper co-operation with the army, a relationship which had existed throughout O'Connor's time and in Crusader but which had been lost when airfields had been overrun in the Gazala battle.

Rommel moved to the attack on the night of 30 August and as expected struck through the southern sector. The battle went exactly as Montgomery had foreseen. In Montgomery's own words he called the tune and Rommel danced to it. In this, his first battle in independent command, he put into practice all about the tactical advantage and the initiative that he had taught at Quetta seven years before. Montgomery fought the battle with caution; he allowed Rommel to batter himself against the Alam el Halfa position, he attacked him with all the air power available, and he harried him with light elements of 7th Armoured Division. But Rommel's army, held in a pocket and badly mauled, was allowed to live to fight another day. Montgomery made a tentative effort to strike south to cut Rommel's strongly held gap through the British forward defences. The attack entrusted to New Zealand Division with a brigade of 44th Division and a regiment of infantry tanks, made little progress and Montgomery decided to keep his force in hand and harry Rommel back through the minefields. Those who belittle Montgomery's victory because he did not finally destroy Rommel forget the number of times that Rommel had triumphed in a battle of opportunity. In order to destroy Rommel Montgomery would have had to move the bulk of his armoured forces from the vital defensive area of Alam el Halfa and this, in the state of training of Eighth Army, he was not prepared to do. Critics forget also the triumph with which Hitler and Mussolini anticipated the advance to Alexandria, the relief with which the outcome of the battle was received in Britain, and, above all, the effect which Montgomery's handling of the battle had on the morale of the Eighth Army and their confidence in its commander.

Critics of Montgomery have also tried to belittle the conduct of the Alam el Halfa battle on the grounds that he fought it on Auchinleck's plan. This is a ridiculous criticism. It is clear from de Guingand, who was Chief of Staff to both, that Montgomery made up his mind independently and without consulting other plans.[18] But even if it were true it would not be a cause for criticism. If a commander takes over a sound plan is he to discard it just because it is somebody else's? Moreover, as has already been pointed out in these pages, any good staff officer can make a sound plan. It is skill in fighting the battle that counts and it is given to few to have this supreme ability in the face of a skilful enemy. No one who has studied the Desert Campaign could believe that Auchinleck would have fought the battle exactly as Montgomery did, and he would be a bold man who would declare that Auchinleck would have brought it to a more successful conclusion.

After Alam el Halfa Rommel saw that there was little hope of inveigling Eighth Army into the kind of mobile battle in which he had so often worsted them. He therefore put his faith in a strongly held infantry position, deeply wired and mined. He now had two Italian armoured divisions and he paired these

[18] de Guingand, *op. cit.*, p. 141.

off with his two panzer divisions, one of each towards each flank, with the task of destroying any penetrating forces before they could achieve a break-through. Montgomery's problem was therefore not unlike that which had faced commanders during the Great War. He had no open flank or room for manoeuvre and he had to blast a hole in the enemy position and to destroy the enemy armour before the hole could be closed.

An important element in Montgomery's plan was to keep Rommel in doubt as to the sector and time of the attack; the only element of surprise that lay open to him. He wanted to keep Rommel's armour separated while he drove a gap through the northern defences and pushed his armour through to an area where Rommel would have to attack it quickly in order to save his infantry from destruction. Instead of the old story it would be Rommel who would have to commit his armour piecemeal against the concentrated British strength. The first phase of the attack was to be carried out by XXX Corps, commanded by Leese; this involved a night attack by four infantry divisions to clear the mines and, supported by infantry tanks, to make the breach. Lumsden's X Corps of two armoured divisions was then to pass through in two wide lanes to the slight ridges beyond Rommel's forward defences. The armour, on ground of their own choosing, would await the attack of the enemy armour while XXX Corps held and widened the gap and destroyed the enemy infantry. Every available artillery unit was to be used to support the infantry advance and the subsequent tank action. The air forces, after their primary task of achieving air superiority were to be used in close support of the battle. In order to deceive the enemy before the attack and to keep his armour separated, all sorts of subterfuges were used including false maps, wireless deception, dummy camps, and above all track discipline. Montgomery thus ensured that all his forces would fight concentrated and with the greatest possible chance of destroying the enemy in detail.

Rommel was away in Austria on the night of 23 October when Montgomery attacked, and did not rejoin his army until the 25th. Even so the battle did not go exactly according to plan; battles seldom do. It took longer than Montgomery had hoped to make the breach in the enemy defences and he had to order Lumsden's two armoured divisions to participate in fighting their way through. Although this possibility had been foreseen in the planning neither Lumsden nor Gatehouse (commander of 10th Armoured Division) liked it; Montgomery had to insist with all his authority before it was done.[19] It is not reading too much into this incident to see in it the working of the old fallacy that armour existed only to fight armour and for the pursuit. Neither Lumsden, the cavalryman, nor Gatehouse, one of the early tank exponents, felt it was right that they should participate in the battle of all arms. X Corps was indeed the *Corps de chasse* and it was intended that it should be kept primarily for that purpose but the main battle had first to be won. Notwithstanding the difficulties and the changes of axis of attack that had to be made before the battle was over, matters went broadly as Montgomery had designed. Above all, Rommel's panzer divisions did break themselves against the armour and infantry that penetrated his defences, and his counter-attacks met the whole force of the Eighth Army artillery.

Montgomery can be criticized for his decision that XXX Corps should make the breach and that the push through for the decisive battle should be under a different corps commander. This accentuated the difficulty in deciding the moment to pass through the armour; it was the type of decision which had so often led to an opportunity being missed during battles of the Great War. Happily Montgomery had been insistent and had chosen the moment. In retrospect it seems that it would have been much better if one corps commander, or two side by side, had had the responsibility for making the breach and for passing the armour through. If the front of attack had been considered too wide for one corps commander to have the full responsibility there would have been difficulties because Horrocks with XIII Corps was engaged in the deception plan in the southern sector and Lumsden was required for the main armoured battle and the pursuit. But these difficulties could have been overcome. The interesting point is that Montgomery did not again use this practice of passing one corps through another. One of the first criticisms he made when he took

[19] Carver, *op. cit.*, p. 135, and de Guingand, *op. cit.*, pp. 199-200.

Tactical Aspects of the War in the Desert 241

over the plan for Overlord was that the whole assault front was commanded by one corps. He altered this so that there were four corps headquarters side by side each responsible both for assault and for follow-up divisions. It is true that at the same time the number of assault divisions was increased from three to five but that was not the reason for the alteration of the command organization.[20]

By 4 November Rommel saw himself to be almost encircled by about twenty times the number of tanks he then had; he decided that he must get clean away with all the mobile forces and in any vehicles he could muster. All made for the coast road and westward, leaving the non-mobile units to get away as best they could. The pursuit was the most disappointing part of Montgomery's battle. The real opportunity to cut off a substantial part of Afrika Korps was lost on the early morning of 5th November. On that day the armoured divisions and New Zealand Division, which had now come under X Corps, were considerably mixed up after heavy fighting and they all found some difficulty in getting under way. When they did start they were directed to objectives which were too close in to catch Rommel. Charing Cross, on the coast road close to Mersa Matruh and one hundred miles from the battlefield, was probably the nearest place where it might have been possible to forestall the bulk of Afrika Korps. 1st Armoured Division was not directed there until the evening of 5 November by which time it had lost time and distance by a short hook north to the coast road. With hindsight it appears that Montgomery's best chance would have been to use 7th Armoured Division for the pursuit. This division had been with XIII Corps in the south and had only moved up for the latter part of the battle. It was therefore fresher and better placed than the other divisions and its commander, Harding, was just the man for the task. But when considering the problem of the pursuit it is well to remember that it is usually possible to get clean away in the desert, as the British had shown on several occasions. The rain which came on 6th November made the pursuit more difficult but the race had been lost before it came.

The greatest disappointment of all came from the air forces whose lack of effectiveness in the pursuit was in marked

[20] Montgomery, *Memoirs* (Collins, 1958), p. 212.

contrast to the magnificent work they had done at Alam el Halfa. On the early morning of 5 November air reconnaissance had reported solid masses of vehicles all the way along the road to Mersa Matruh. The Desert Air Force had been specially prepared for participation in the pursuit and as the Eighth Army advanced they expected to see the sides of the road littered with wrecked vehicles; in fact there were hardly any. For some reason the air attack had been delivered by high level bombing which was singularly ineffective for such a purpose.

Montgomery was well aware that, as one officer told him, the Desert Army had been in the habit of going up to Benghazi for Christmas and getting back early in the New Year.[21] He was going to make quite sure that this time Rommel was not going to be allowed to come back at El Agheila. He ensured that light forces were pushed right through the gap there to hold the narrow end of the funnel until he was ready to continue his advance. This precaution had not been taken either by Neame in 1941, because of his small and immobile force, or by Ritchie in January 1942. Rommel then was kept on the run and made his next stand at the Mareth line on the frontier of Tunis 1,100 miles from El Alamein. By that time the threat arising out of the Torch landings was beginning to develop and the days of the Axis in Africa were numbered.

[21] Montgomery, *op. cit.*, p. 140.

Chapter 12

TACTICAL ASPECTS OF THE WAR IN EUROPE

The lessons brought out in the great combined operation for the capture of Sicily were to stand the British in good stead in the later vital landing in Normandy. The lessons were to a large extent technical; in particular the Allies learnt how to handle the array of ships and aircraft necessary for such an operation. In the landings and in the subsequent battles Montgomery showed his respect for the principle of concentration. He planned for landings within supporting distance of each other and always wanted a battle under the single control of one commander. The Americans, and indeed some of the British planners preferred wider converging movements and often thought him too cautious. Alexander, for the second time commanding an Allied operation, controlled the battle with a lighter rein than Montgomery. This had advantages in dealing with an ally but since Montgomery always knew what he wanted to do and lost no opportunity of doing it, it sometimes had disadvantages too. In Sicily both Patton (Seventh Army) and Bradley (II Corps) felt that Montgomery had been allowed too much of his own way to the detriment of what United States forces could have achieved.[1]

It would be easy to make too much of the rivalries and jealousies of the Allies especially as the British and Americans worked together more closely and with less friction than any major allies in history. As will be seen later Bradley worked loyally under Montgomery when he was formally under his command in Normandy. Perhaps the Sicily pin-pricks were remembered when Montgomery and Bradley held equivalent positions as army group commanders and then Eisenhower had to be careful not to give Montgomery all his own way. In that later campaign Bradley had the headstrong Patton under his command, a position reversed from that in Sicily.

If the Italian campaign has any lesson it is that no amount of military genius and no amount of heavy equipment can solve

[1] Jackson, *The Battle for Italy* (Batsford, 1967), p. 78.

the problem of the attack in confined areas and bad weather. The Gustav line from Gaeta through Cassino to the hills north of the River Sangro defied alike the carefully prepared attack of Montgomery's Eighth Army and the more slapdash methods of Mark Clark's Fifth Army. Moreover in the difficult country of southern and central Italy, where roads were few, neither the British nor the American armoured formations made any real impression even in good weather.

The geography of Italy suggests amphibious operations. Apart from the shortage of landing craft,[2] the difficulty was so to place and to time the landings that they would not be erased before supporting forces labouring through difficult country could link up with them. The Anzio landing was timed perfectly in that it went in just at the moment when Kesselring had committed his last reserves against Fifth Army on the Garigliano and at Cassino. Nevertheless the Allied link up did not take place until almost five months later. The Allied inability to take advantage of a successful landing was partly a failure in command but it owed much to the fact that the main operations around Cassino were bogged down in almost impossible conditions of terrain and weather.

The operations at Anzio and the subsequent attempt to destroy Kesselring's army south of Rome both illustrate the difficulties of Allied command. Anzio shows Alexander, Clark (the army commander), and Lucas (the corps commander), all looking at the battle in a different way and it also shows Alexander's inability to make the two Americans fight the battle his way. Alexander may have been inspired by Churchill's enthusiasm and his desire to capture Rome as soon as possible, but the object of the Anzio operations and the fact that it was subsidiary to the Garigliano and Cassino attack is clear from his instructions. He gave the object as: ' of cutting the enemy lines of communication and threatening the rear of the German XIV Corps.' He went on:

> The enemy will be compelled to react to the threat of his communications and rear, and advantage must be taken of this to break through his main defences and to ensure that the two forces operating under Commander Fifth Army join

[2] See Ch. 10.

Tactical Aspects of the War in Europe   245

hands at the earliest possible moment. Once this junction has been effected Commander Fifth Army will continue the advance North of Rome with the utmost possible speed. . . .[3]

Clark's instructions to Lucas rightly concentrated on the necessity of obtaining a firm hold on the Anzio beaches. What was not so laudable was the vagueness with which his subsequent actions were directed. Lucas's second ' mission ' was described as ' advance on Colli Laziali '; these are the Alban Hills, south-east of Rome, which lie twenty-five miles inland from Anzio and had been specified by Alexander as a suitable objective for Lucas's corps. Clark's ' advance on ' could mean ' in the direction of ' or it could mean that the hills were the objective. That this vagueness was deliberate is shown by the oral elaboration of his instructions given to Lucas by Clark's chief of operations staff. Brigadier-General Brann explained that all that was required of Lucas was the securing of the beachhead; the wording of the order was deliberately vague so that Lucas would not feel bound to risk his corps in a rash advance to the Alban Hills.[4]

Clark was thinking not only of the effect of Lucas's landing on his own main operation but of the German reaction to the landing. In view of its proximity to Rome and important German airfields he believed the Germans would be bound to concentrate on the destruction of the beachhead and that Lucas's essential task was to guard against this possibility. It is small wonder therefore that Lucas concentrated on consolidation of the beachhead and regarded the advance to the Alban Hills as very much a secondary operation which might be possible later. Both Alexander and Clark visited Lucas in the beachhead on D Day, 22 January, when it was clear surprise had been achieved. Neither of them seems to have taken steps to ascertain Lucas's future intentions and both appeared satisfied. Clark's last words to Lucas were: ' Don't stick your neck out, Johnny, I did at Salerno and got into trouble.'[5]

Kesselring's reactions were immediate and were taken with-

[3] 15 Army Group Instruction, No. 34, 12 Jan. 1944.
[4] *Command Decisions*, p. 257.
[5] *Ibid*. p. 262.

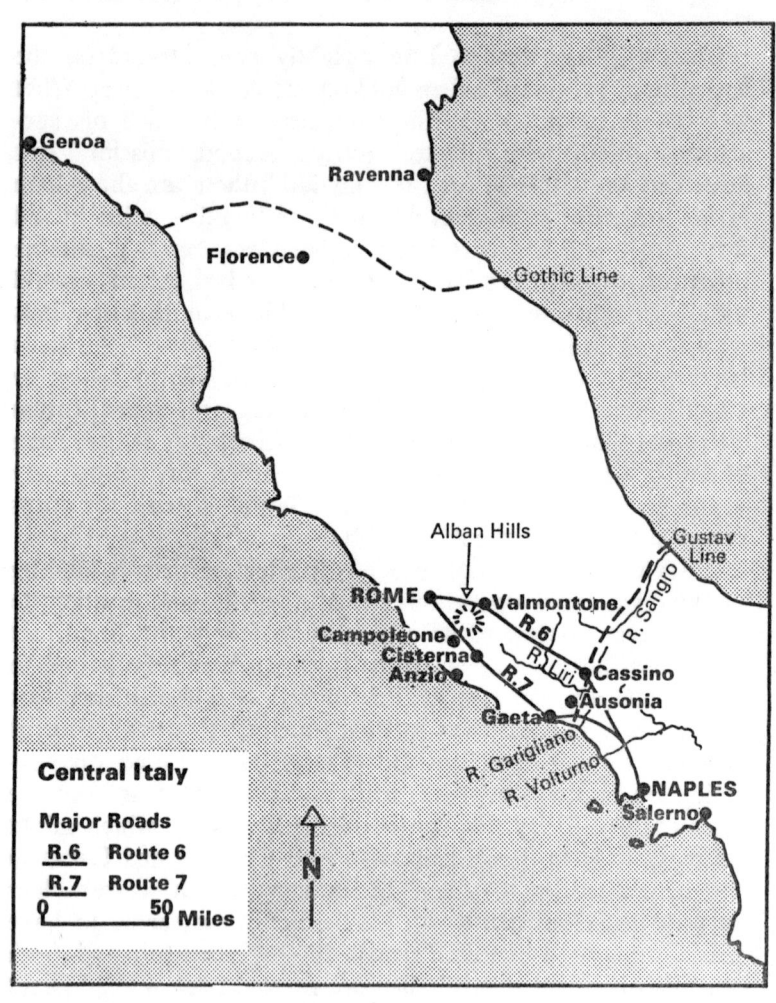

Map 11.

out affecting the German resistance to the main Fifth Army attack. He brought in divisions from the Adriatic coastal sector and brought down Mackenson's Fourteenth Army Headquarters from the north with some of his divisions; so that Tenth Army would be free to deal with the main battle. By 25 January, the day when Lucas first made an attempt to extend his beachhead, the Germans were holding the exits at Campoleone and Cisterna. On that day Alexander and Clark again visited Lucas but it was not until 27th that Alexander appears to have expressed to Clark his dissatisfaction at the slow progress. Alexander then pressed for the capture of the line Campoleone-Cisterna, a course that might well have been pressed on him on the first day. Clark went back to Anzio to urge Lucas to take this action. By 29 January Lucas was ready with a small reserve in hand and he was even thinking of getting to Valmontone and to the control of Route 6 on which the German Tenth Army depended. But it was too late; neither 1st British nor 3rd United States Division made any headway. The American 'Rangers', the equivalent of British Commandos, suffered a disaster and lost two out of three battalions in an ambush and the United States 1st Armoured Division, committed in unsuitable country, achieved nothing. Whatever else may be said about the conduct of the operations in the beachhead the lack of early reconnaissance which made this mistake possible cannot be forgiven.

The Allies did not know how close their attack was to success, but after its failure it was their turn to withstand an attack by the Germans in which eventually ten divisions were employed.[6] To their eternal credit Lucas's five divisions held on to the minimum defensive area around the port of Anzio. Despite the Germans' local air superiority the port and the essential administrative depots continued to work and it was possible to arrange the relief of some of the battered divisions.[7] This magnificent defence and the continued defence of the beachhead against an enemy that overlooked the Allied positions almost everywhere, ensured that, despite the disappointments, the Anzio beachhead was able to pay its dividend when a few months later the final battle for Rome took place.

[6] Jackson, op. cit., p. 200.
[7] The author was commanding one of the relieving battalions and took over from a battalion of 56th Division about seventy strong.

Lucas has been blamed for the early failures at Anzio. Certainly his diaries[8] show a man who was tired and lacking in the faith and energy for such a venture. But diaries can be misleading and, if he was not the man for the job, Clark should have replaced him of his own accord, and if not, Alexander should have insisted. Here the difficulty is that of an officer of one nation dealing with the principal commander of the army of another nation. What would the British have said if Eisenhower had insisted on Montgomery removing one of his principal subordinates? Commenting on this aspect Cyril Falls says:

> British soldiers have reproached the Supreme Commander in North-West Europe, General Eisenhower, with a tendency to act as the chairman of a board rather than as a supreme commander... It would seem, however, that the Americans hold material for a powerful retort to the criticism of General Eisenhower. They might well ask whether chairmanship was in fact better exemplified in North-West Europe than on the western side of the Apennines.[9]

Before condemning Lucas it is worth considering what might have happened if he had tried to fulfil Alexander's intention of getting to the Alban Hills before he had fully established his beachhead. In the light of events as they happened it seems almost certain that in the subsequent German counter-attack his force would have been cut off from Anzio and destroyed. Lucas's mistake was not in failing to press on to the Alban Hills but in failing to establish his defences far enough out. If he had taken advantage of the surprise he achieved in the landing he could without doubt have secured Campoleone and Cisterna and, despite the extra front involved, this would have made his defensive task easier. Such a forward defence should have been accompanied by widespread raids and fighting patrols sent out towards Rome and the German communications. This is the kind of action that Alexander and Clark could have indicated when they

---

[8] Extensively quoted in *Command Decisions*, Ch. 11, which is an illuminating account of the operations from an American point of view.
[9] *Command Decisions*, Introduction (English edition only), p. xiii.

Tactical Aspects of the War in Europe 249

visited Lucas on the day of the landing. Such advice could well have been tendered without fear that they were interfering in Lucas's conduct of his own battle. Here a comparison with Montgomery's command methods is apposite. One of the ways in which Montgomery showed his genius in the tactical battle was that he was always able to make sure all the way down the chain of command that the corps or other commander, right down to the battalion commander, who had the important task to do, knew what his part in the battle was and what was expected of him.

Alexander's next battle was a masterpiece. By a series of attacks at Cassino, in the Liri valley, in the difficult hill country further west and by the break out from the Anzio Beachhead, he severely defeated the German Tenth Army and captured Rome. There were two principal reasons why the battle did not lead to even more decisive results. The first was that the Tenth Army escaped destruction because Clark did not obey Alexander's instructions to the Fifth Army. The other was that the pursuit was prejudiced by the withdrawal of divisions to take part in Operation Dragoon.[10]

The breaking of the Gustav Line was achieved by the concentrated blows by Eighth Army, which had been secretly concentrated west of the Apennines, and Fifth Army, which included Juin's French Corps. By masterly deception methods the attack came as a complete surprise. The Eighth Army attack at Cassino and in the Liri Valley was more successful than the three earlier attacks, partly because it was the first one to be delivered in suitable weather, in sufficient strength, and on a wide enough front. The real reason, however, for the advance in the Liri Valley and in the American coastal sector was the magnificent breakthrough by the French Corps in the battle in the mountainous sector between them, and their capture of Ausonia. On the other flank the two Polish divisions, by a magnificent feat, captured Monastry Hill, which had defied so many previous attacks.

Alexander, who had kept in his own hand the timings of the breakout from Anzio, indicated clearly how the two armies were to exploit this success. The Eighth Army and the main body of the Fifth Army were to continue the advance

[10] See Ch. 10.

in the direction of Rome while VI Corps from Anzio struck across to Valmontone on route 6 to cut the main axis of retreat of the German Tenth Army.[11] Clark did not fully accept the idea that the only proper course for VI Corps was to make for Valmontone and had, without telling Alexander, instructed Truscott, its commander, also to consider the possibility of a direct advance to Rome. On 17 May, as the moment for the break-out was approaching, Clark did discuss with Alexander the difficulty of an advance to Valmontone and suggested that a decision might wait until the enemy situation and reactions were better known. Alexander was adamant and made it clear that Valmontone was the objective and indicated the manner in which the battle there might be fought. Notwithstanding this, after VI Corps succeeded in winning the battle for Cisterna, Clark directed Truscott to make his main thrust with four divisions to Rome and only a secondary move with two divisions towards Valmontone. The result of this dispersal was that neither objective was immediately attained. Alexander was told of the deviation from his orders after it was too late to alter it; he accepted the action with a surprisingly good grace. In the end Rome was captured from the direction of Valmontone but by that time much of the defeated Tenth Army had slipped away. Clark seems to have been genuine in some of his fears that VI Corps would not be strong enough to bar the retreat of Tenth Army, but there is little doubt that his real reason for the deviation from orders was that he was attracted more by the prospect of the capture of Rome than the destruction of the enemy and that he wanted to make certain that United States troops were the first to enter Rome. Had he obeyed Alexander's orders his aim could have been achieved more quickly and the victory would have been more complete.[12] By his departure from Alexander's orders Clark took out the corner stone from the plan. It is true that there were several minor roads and tracks passing east of Rome along which part of Tenth Army did escape. Nevertheless Valmontone was an essential road centre and the magnificent concentrated attack made by the two American

---

[11] Allied Army Italy, Operation Order No. 1, dated 5 May 1944.
[12] *Command Decisions*, Ch. 12, gives a full account of Fifth Army in these operations.

Tactical Aspects of the War in Europe 251

corps of Fifth Army after the first checks at Valmontone and on the road to Rome[13] suggest that if the Fifth Army had been fought according to Alexander's orders the Tenth Army might well have been destroyed.

Before passing on to the Normandy operations it is worth looking at two aspects of the fighting in Italy, the contribution of the air forces and the performance of a highly mechanized army in formidable terrain. During the campaign the Allies gradually built up a degree of air superiority which led some to believe that air forces could win the battle for the soldiers on the ground. This was a mistake similar to the idea prevalent in 1916 and 1917 that artillery could conquer and that infantry only had to occupy. The extreme view was held by airmen who believed that air forces could break the deadlock south of Rome by bombing the enemy communications, starving the Tenth Army of all the necessities of battle and bombing out of existence the many strong points such as Cassino.[14] An example of air support which suffices to show the fallacy was the destruction of Cassino and the Monastery in the unsuccessful attacks of February 1944. All the bombing achieved was to make the task of the German defenders easier. The cult of air support also had a dangerous by-product. There grew up a tendency to regard an attack as impossible unless air support could be provided and in defence it became more natural to call for air support than to use the weapons that were in the soldiers' hands. This is not to belittle the part that air forces can play in battle. The first task of air forces is to gain air superiority and the army that works under that umbrella has a freedom of action, a facility to deceive the enemy, and an administrative advantage that is denied to the enemy. But direct air support is just one more weapon to be used as part of the armoury. By itself it can achieve nothing.

The poor showing of armour in the Italian winter has already been mentioned. In the battle for Rome the weather was dry and the going good for the tanks, yet the British and Canadian armoured divisions attacking side by side in the Liri Valley made little impression on the enemy. The French Corps, with few vehicles and depending chiefly on mule transport, made

[13] Jackson, *op. cit.*, pp. 245-246.
[14] Plan 'Strangle', see Jackson, *op. cit.*, p. 205.

quicker progress whether it was fighting in difficult mountain country or was taking advantage of ground to outmanoeuvre the enemy. In the American sector too it was the infantry and not the armour that played the major part. Nor was it only the armoured divisions that were slow; it is not too much of an exaggeration to say that the mobility of a division was in inverse ratio to the number of vehicles it took into battle. In order to fight in Italy men had to attack and defend in areas where they could only be supplied by mules or by porters. The change in conditions from the quick dash forward from the toe of Italy to the hard fighting in front of the Gustav Line, followed by the pursuit and the advance into Northern Italy, demanded a flexibility of organization that was not required either in North Africa or in North-West Europe.

For the assault on Normandy the Allies used every possible contrivance and every stratagem for deception that their combined effort could devise. The soldiers in high places showed none of the reluctance to accept and foster new ideas that has been attributed to their elders of the Great War. Hobart was given a special command, known as 79th Armoured Division, to develop and use weapons and machines to overcome beach defences, mines and obstacles. Admiral Mountbatten, Chief of Combined Operations before he went to South East Asia, had been energetically dealing with the problem of landing craft and under-water obstacles. The administrative problem was faced in the same way and resulted in the invention of the Mulberry Harbour and Pluto—the Pipe Line Under The Ocean. These subjects and the Royal Air Force use of radar and other devices in furthering the deception plan are outside the scope of this study; it suffices to say that no attitude of obscurantism or conservatism was allowed to stand in the way of their successful use in Overlord.

Montgomery was in command of the land operations until such time as Eisenhower, the Supreme Allied Commander, decided to take over command himself. This change in command would be required when the beachhead operations were completed and both the British 21st Army Group and the American 12th Army Group were deployed. Montgomery in the meantime had under his direct command Dempsey commanding the British Second Army and Bradley commanding

Tactical Aspects of the War in Europe 253

the United States First Army. Montgomery's plan was that after the capture of the beaches a lodgement area was to be established from the River Orne east of Caen to about Avranches. The importance of this lodgement area was that it contained the port of Cherbourg and airfields and that the possession of Caen gave the Allies a hold on roads and railways pointing into the heart of France. By posing a direct threat which the Germans could not ignore Mongomery intended to draw the main German strength, particularly their armour, against his left or eastern sector. This would set the scene for a break-out by the Americans on the western flank in a wide sweep to the Seine south of Paris. As at El Alamein the battle did not go exactly according to plan but in all essentials Montgomery achieved his aim.

Included in the orders and instructions for the battle was a phase map which showed that Caen should be captured by D plus five (11 June) and Avranches and Falaise fifteen days later. It was estimated that the Seine would be reached in three months. In fact, although the landing achieved surprise and the beaches were secured according to plan, Caen was not finally cleared of the enemy until 20 July. Nevertheless the Seine was reached a week ahead of schedule. The point to remember about the phase table is that it is of no importance except as a guide to the administrative staffs. Montgomery had to be free to fight the battle according to the situation as he saw it, not according to some rigid time-table—even if he himself had been the author of the time-table. It is to Montgomery's everlasting credit that he refused to be disturbed by other people's doubts and difficulties. From the moment of landing Montgomery never lost the initiative and he gradually imposed his will upon the enemy with the result that they suffered such a defeat in Normandy that they were unable again to offer battle until they were close to the borders of Holland and Germany.

Although now that we can judge the battle as a whole Montgomery's battle strategy can be seen to be right, it was not so clear at the time. At Eisenhower's Headquarters there was considerable dissatisfaction with progress and the press, particularly in the United States, voiced its doubts. Because of this the doubts were represented as an Anglo-American row. It was nothing of the sort; Eisenhower's Headquarters was an Allied Headquarters and most of the critics were senior British officers. The strongest critic, and indeed by his own showing the originator of most of the criticisms, was Tedder who was Deputy to Eisenhower and thus the senior British officer at the Headquarters. The descriptions by Eisenhower and Tedder in their respective books[15] hardly read like the same operations. Tedder certainly had cause to urge the expansion of the beachhead because without this there was not sufficient room to establish the airfields so urgently required. Eisenhower did go so far as to express his fears to Churchill that Montgomery was being unduly cautious and this made Brooke think Eisenhower did not fully understand Montgomery's plan.[16] It is true that Eisenhower did not seem to think that the battle was being waged to the full unless everybody was attacking all along the line.[17] Nevertheless Eisenhower was at his best in seeing that doubts and differences were not allowed to interfere with the conduct of the battle; supported by Eisenhower from above and Bradley from below Montgomery was allowed to fight the battle to its successful conclusion.

It was the bitter fighting about Caen that caused most of the doubts about Montgomery's handling of the operations. The battle from 18 to 21 July, known as Goodwood, is worthy of special mention.[18] The object of the battle was to get VIII Corps of three armoured divisions out on to the high ground some six miles south-east of Caen with possible exploitation towards Falaise. One of the principal problems was to get this large force, some 750 tanks, through a small salient in which the River Orne and a railway embankment had to be crossed. Surprise was difficult because the enemy overlooked the salient from the high ground north-east of Caen. By a carefully planned air and artillery bombardment the task of getting the tanks across the river and over the railway was accomplished and first objectives were captured before the

---

[15] Eisenhower, *Crusade in Europe* (Heinemann, 1948), pp. 291-294. Tedder, *With Prejudice* (Cassell, 1966), pp. 553-563, 569-574
[16] Bryant, *op. cit.*, vol. II, pp. 243-245.
[17] Eisenhower's report to Combined Chiefs of Staff dated 13 July 1945, quoted in Montgomery, *op. cit.*, p. 255. *Ibid.* p. 262.
[18] Ellis. *op. cit.*, pp. 335-350.

Tactical Aspects of the War in Europe 255

enemy recovered from the bombardment. After that, by the use of anti-tank guns and tanks the enemy were able to inflict heavy tank casualties and bring the attack to a halt. British difficulties were increased by heavy and prolonged rain on 20 July.[19] This battle is another example of the very mass of the armoured attack defeating its own object. The proportion of infantry to armour was too small for an integrated battle and the mass of tanks and vehicles was such that the artillery could not get forward to support the subsequent phases of the attack. Much reliance was placed on air support, especially anti-tank rockets, during this phase but this was much hampered by low cloud. Nevertheless, Rommel, commanding the Army Group in this his last battle, had to use most of his armoured resources to halt the attack. By this battle therefore Montgomery did achieve his object of containing the enemy armour and did facilitate the great American breakthrough which was to reap the fruits of victory. A measure of his success is that when a few days later the American attack went in there were 190 German tanks in their sector and 654 in front of the British. Most of the 88 millimetre anti-tank guns also were massed against the British.[20]

By the end of July the Americans had captured Avranches. By this time there were two Allied Army Groups in Normandy, Montgomery's 21st consisting of First Canadian and Second British Armies and Bradley's 12th made up of First and Third United States Armies. Hard fighting continued in the first week of August; the Canadians were trying to get to Falaise while the Americans, supported on their left by the British, were trying to punch a hole to get Patton's army through. At this stage Hitler took a hand and ordered Kluge (who had replaced Rommel and also superseded Rundstedt as Commander-in-Chief West) to strike with all his armour to Mortain and the coast north of Avranches. This attack came in too late to be effective. Patton was already breaking through and Bradley was easily able to contain the German armour, against which air attack wrought havoc. The only result of Hitler's intervention was to put most of his force into the sack just as we were in a position to close the mouth. Although Mont-

---

[19] Belfield & Essame, *The Battle for Normandy* (Batsford, 1965), pp. 134-145.
[20] *Ibid*. pp. 179-180.

gomery was still directing both Army Groups the build-up had reached the stage at which Eisenhower intended to take over operational command. He was looking for the opportune moment and began increasingly to intervene in the battle. Bradley was beginning to make decisions on his own or bypass Montgomery direct to Eisenhower. Thus for the first time in the Normandy battles Montgomery did not have complete control. The mouth of the sack was not completely closed, but this was not altogether because of divided control nor was it altogether a bad thing. The Canadians made slow progress to Falaise and the Americans who reached Argentan, their objective, did not press on north to meet them. There was thus a gap of twenty-five miles through which the Germans raced to escape; in this attempt their converging columns were subject to remorseless and effective attacks from the air and artillery. The Allies on the other hand avoided the complications of a head on meeting much further north than contemplated by the original orders.[21] The German casualties are difficult to assess but they lost more than 10,000 killed and about 50,000 prisoners. In the Normandy battle as a whole German casualties were over 300,000 and those of the Allies almost 210,000 (including 37,000 killed).

The air forces played a tremendous role in the battle for Normandy. The landing could never have taken place at all had it not been for the degree of air superiority achieved, but perhaps their most important contribution was their share in the deception plan which ensured that the Fifteenth Army, about half Rommel's Army Group, was out of the battle until late July, because it was watching the Calais area. Once the landing had been achieved much of the air effort was directed to interdiction, so making the movement of German reserves almost impossible. It was the knowledge that this would happen that had made Rommel insistent that the German effort should be directed to preventing landings and to erasing those that got a foothold. Rundstedt, Commander-in-Chief West, would have preferred the beaches to be more lightly held and a strong central reserve retained in hand to destroy the invaders before they consolidated their beachhead. Hitler had overruled Rundstedt without giving Rommel control of all the reserves. In

[21] *Command Decisions*, Ch. 14.

Tactical Aspects of the War in Europe 257

addition to their strategic tasks the air forces provided support on a lavish scale for the land battle. Great attention had been paid to technique and procedure during the training period and it was even found possible on occasions to use the heavy bombers in direct support. On the whole, and in the early stages particularly, the army were not sufficiently selective in choosing the targets and asked the air force to do too much. As in Italy they were inclined to overlook the problems which bomb damage raised for the attackers. What happened at Cassino happened again at Caen.[22] For Goodwood the technique was much better but it was in the destruction of the German counter-attack at Mortain and then the destruction of so much of the army making its way back through the Falaise gap that the air forces really came into their own.

Eisenhower did not take over operational command until 1 September, but on 24 August, the day before Paris fell, he issued a directive to both army groups.[23] This instructed 21st Army Group to seize the Pas de Calais and the airfields in Belgium and to establish a firm base at Antwerp. Twelfth Army Group was ordered to clear Brittany and make for Metz. Montgomery did not approve of this plan nor of Eisenhower taking over operational control. Even if he could not get his way over command he thought that he should be given the task of striking straight for the Rhine. He believed that the enemy would have to concentrate to defend the Ruhr and that their defeat would lead to the end of the war in 1944. Montgomery realized that the fifteen divisions of 21st Army Group would not suffice to achieve this and to make his plan feasible he required under his command an American army of at least twelve divisions. Montgomery's appreciation of the situation was based on the belief that only a single mind directing the pursuit could enable the Allies to fight the next battle before the Germans had time to recover. Moreover the administrative problem of the move across half Europe and a battle on the Rhine was such that it could only be solved if all resources were devoted to the needs of the commander fighting the battle.

[22] de Guingand, *op. cit.*, pp. 401-404.
[23] Confirmed in a letter received by Montgomery on 29 August reproduced in Ellis, *op. cit.*, pp. 474-475.
See also *Grand Strategy*, vol. V, p. 381.

17

Eisenhower saw the battle very differently. He believed that the enemy would strain every nerve to hold or at least delay us in their strong defences on the Siegfried Line. The Allies must therefore follow up as quickly and on as wide a front as possible. They would thus be in the best position to find out where the enemy was weakest and would themselves have available the strongest possible force to strike the final blow. His course would have the advantage of linking up the advance from Normandy with that from the Dragoon landing in the south of France. Eisenhower could not contemplate allowing Montgomery to continue in operational command especially as Montgomery wished to combine this with command of 21st Army Group. Apart from the relative soundness of the two tactical concepts Eisenhower's views on command were easily understandable. The British and Canadians would be unlikely to increase beyond the fifteen divisions then in action; the Americans had twenty divisions on 1st September and many coming forward. They had fifty-six in October and eighty-one by February 1945. With their growing forces it is not likely that the United States forces could be subordinated to a British commander for long. In answer to this Montgomery expressed himself as willing to serve under Bradley in a single thrust if that were preferred.

It is conceivable that Montgomery, with his real understanding of the position and his mastery of battle technique, might have succeeded if he had been given what he asked. Most of the German commanders who opposed him have said that Montgomery was right. On the other hand in the cold light of reason Eisenhower's looks the sounder plan and there were tactical and administrative arguments to support it. It is relevant that Montgomery's chief of staff, de Guingand, differed from his commander on this subject, the only major issue in which he ever did.[24] There was wisdom in Montgomery's contention that maintenance is easiest when the staff can work to a simple plan in which they know exactly what the commander fighting the battle wants and when all resources are at his disposal. But the advance on a wide front allows the maximum number of roads, ports and railways to be used and facilitates air supply. From the tactical point of view the

[24] de Guingand, *op. cit.*, pp. 411-412.

single thrust, strong though Montgomery intended it to be, gave the Germans the opportunity for delaying operations which would have given them time to recover. The German reaction to the Arnhem operation is an example of their ability in this direction and the Ardennes offensive shows their genius for an unexpected riposte.

Whether Eisenhower or Montgomery was right there is no doubt that Montgomery made one big mistake, possibly his one major mistake during all his time in command. This was his failure to ensure full control of the Scheldt waterway to Antwerp. Antwerp was captured on 4 September. No attempt was made to push on to Breda, which would have cut off troops in Walcheren and South Beveland and, if other operations had been foregone, enabled the garrisons of these islands to be mopped up. As it was large numbers from Fifteenth Army from the Pas de Calais escaped over the Scheldt to swell the defending force north of the river. When, after the failure of Arnhem, Montgomery turned to clear the north bank he found it a very hard and costly operation.

Montgomery was right to put his strongly held views on the future conduct of operations to Eisenhower. It is even more certain that once he had got Eisenhower's decision he should have ceased concerning himself with the broad strategy of the campaign and got on with the tasks he had been given. His attitude should have been guided by his own simple aphorism which he had so often used at the Staff College Quetta: 'Always protest but when the decision is given do not bellyache.' Single minded consideration of the task given him would have shown Montgomery that no powerful thrust into Germany could be maintained without the port of Antwerp. On 7 September Montgomery received fresh instructions from Eisenhower which mentioned the necessity for ' opening the ports of Havre and Antwerp, which are essential to sustain a powerful thrust deep into Germany.'[25] By this time Montgomery realized his administrative situation would not allow him to fight a major battle for the Ruhr; nevertheless, with Eisenhower's support, he turned not to the clearing of the Scheldt but to the Arnhem operation.

One of the reasons for the Arnhem operation was to get

[25] Montgomery, *op. cit.*, p. 273.

British forces into Holland to capture the rocket sites (V 2) which had come into action against London. Montgomery believed that this operation would draw German troops away from the Scheldt and facilitate the opening of Antwerp. Another of the reasons for Arnhem was not tactical at all but arose out of the desire to use First Allied Airborne Army. This fully trained force (three United States and two British airborne divisions) was ready in England and 'was burning holes in SHAEF's[26] pocket'. It was not intended to use the airborne troops in any rash manner but Eisenhower's staff 'had decided to buy an airborne product and were shopping around'.[27] It had been intended to use airborne troops in the pursuit across France but the land forces had moved too fast and aircraft had been more profitably employed in supply duties. Of eighteen airborne operations planned three possibilities now remained: a landing in Walcheren to clear the Scheldt; an operation to seize a crossing over the Rhine at Wesel; and the Arnhem operation. The first was rejected because of the ease with which the Germans could flood Walcheren. Wesel was an objective much favoured by Dempsey, commander of Second Army, because it would bring him a Rhine crossing close to 12th Army Group's area of operations. Montgomery chose Arnhem because of air force advice that Wesel was too strong in anti-aircraft defences. Another air consideration was that the three bridges near Grave, Nijmegen and Arnhem were in an alignment comfortably within range of United Kingdom bases.

The plan for Arnhem was bold but, in the light of the Intelligence estimate of enemy forces made at SHAEF, not unsound. It is certainly an answer to those critics who think that Montgomery was always too cautious. Montgomery had been allotted I Allied Airborne Corps under Browning, consisting of one British and two United States airborne divisions. This corps was put under command of Second Army from which XXX Corps was given the task of racing forward to link up with the landings while its other two corps worked forward on each flank. 1st Airborne Division was given the task of capturing the bridge at Arnhem, sixty miles north of

[26] Supreme Headquarters Allied Expeditionary Force.
[27] *Command Decisions*, p. 335.

XXX Corps forward positions. The 82nd United States Airborne Division was to capture Grave, twenty miles south of Arnhem and to go on to capture Nijmegen half way between the two. The 101st United States Airborne Division was to land between XXX Corps and Grave to deal with two canal crossings. In the end all the bridges were secured except at Arnhem where only one end of the railway bridge was captured. XXX Corps fighting on a narrow front, indeed little more than a road running through sixty miles of enemy-held country, failed to get through. After a valiant fight in which 2nd Parachute Battalion, commanded by Lieutenant-Colonel Frost, denied the enemy the use of the bridge for four days, the attempt at Arnhem was called off and such of 1st Airborne Division as could, ferried themselves across the river and broke out as best they could to XXX Corps positions.

There were a number of reasons for the failure at Arnhem. Bad weather, which hampered reinforcement and supply by air, played its part but the speed and skill with which the Germans reacted was the principal cause. The German XI SS Panzer Korps was in the Arnhem area, refitting after the Falaise battle. Model, now the Army Group commander, was near Arnhem and saw the parachute drop. Student, commander of First Airborne Army near Eindhoven where they could act against XXX Corps advance, had become fully aware of British plans by the capture of the Allied orders from the body of the American airman shot down. The speed and determination shown by the Germans was such that within a few hours they were in a position to isolate and frustrate the Arnhem landing and to bar progress to XXX Corps advance. The optimistic belief that the Germans were incapable of further serious resistance, a belief held in intelligence circles both at SHAEF and at headquarters 21st Army Group, was thus shown to be unwarranted.

The German ability to react to the threat adds weight to the belief that Eisenhower's policy of pursuit on a wide front was right and certainly to the requirement for freeing Antwerp as a base for further operations. Leaving this aside, there remains the question whether the whole airborne operation could have been better managed. Montgomery considers that he got insufficient support from Eisenhower, who had assumed general control of operations. Certainly Bradley's army group was so far away and so dispersed as to leave the Germans free to press at will against the sixty mile salient at Arnhem. At Arnhem itself 1st Airborne Division dropped and landed too far from their objective. The reason was to avoid anti-aircraft fire but the result was the waste of several hours of vital time. All this leads to the thought that this was another of those battles that lacked a single controlling mind. An airborne corps operation, worked out in London, was grafted on to a land advance which had not been properly worked out. Arnhem was a Second Army battle but none of the accounts of it give it the imprint of being Dempsey's battle, nor does it read like Montgomery's battle.

By December the Allied administrative situation had much improved because Antwerp, Marseilles, and a number of the Channel ports were working. Both Eisenhower and Montgomery were preparing for a battle for the Rhineland, confident that the Germans were in no position to launch an offensive, when the counter-stroke in the Ardennes came. Rundstedt, who had returned to command in the west, used thirty divisions in an ambitious effort with the ultimate object of the capture of Antwerp and the separation of the British and American Armies. The attack came in a sector of Bradley's army group which he was intentionally holding only lightly. Since this was largely an American battle it need not be studied in detail here. However, because of the difficulties in command caused by the successes of the German thrust two of Bradley's three armies, First and Ninth, which were north of the penetration, were put under Montgomery's command. Montgomery handled the situation with his usual calm and skill. British forces were moved to ensure a hold on the River Meuse and the American Armies were directed to certain withdrawals which facilitated the gradual establishment of a firm defence while one corps was prepared for the counter-attack. Montgomery was able to put into practice one of his oft-enunciated teachings that in a fluid defensive battle the commander must make up his mind where he intends to hold the enemy, must not fritter away forces in operations irrelevant to this purpose, and that the situation must be stabilized before a major counter-attack is launched. The counter-attack when it comes must be as well prepared and supported as any other attack.

After the situation had been stabilized Montgomery showed, not for the first time, that he was less adept at handling a press conference than in fighting a battle. What he said appeared to patronize the American commanders and soldiers. They knew they had been caught napping and that by magnificent fighting under American commanders in the south and Montgomery in the north they had retrieved their mistakes. There was no more to be said. Up to this time it had been easy to scoff at Montgomery's eccentricities and to delight in rivalling his exploits in battle. Nevertheless to most responsible United States officers and men Montgomery was a bit of a legend and many of his admirers were deeply hurt by what he said on this occasion. It says much for Eisenhower's generosity of mind and truly Allied outlook that he did not allow this incident to affect him and that he again put an American army under Montgomery's command for the Rhineland operations and the crossing of the Rhine.

It had been intended to fight the Rhineland battle in early January when the ground was frozen and movement off roads possible. The Ardennes attack prevented this and delayed the offensive until the February thaw. Twenty-first Army Group was responsible for clearing the west bank of the Rhine up to Düsseldorf. For this task Montgomery planned a converging attack. First Canadian Army (one British and one Canadian armoured division and a Canadian armoured brigade and six British and three Canadian infantry divisions) was to attack through the Reichswald towards Wesel and Ninth United States Army was to strike north from the River Roer with its left directed on Wesel. Second British Army remained free to prepare for and organize the crossing of the Rhine but for this attack they would have to use many of the divisions taking part in the Reichswald battle.

The country which the Canadian attack had to cross was a wedge with its point at Nijmegen running forty miles between the Rhine and the Meuse. The country lent itself to defence and there were three prepared positions in depth, including parts of the Siegfried Line. For the attackers there were only two roads forward and both were close to the rivers and flooded area. At the moment of attack the enemy had only two divisions but before Ninth Army, which was delayed by the opening of the Roer dams, could begin their advance there were eighteen enemy divisions engaging First Canadian Army. The result of this was that when on 23 February the Ninth Army attack did go in they found a situation very much to their advantage. By 3rd March they had linked up with the Canadians and together they badly mauled the eighteen German divisions and inflicted some 90,000 casualties.

A number of special vehicles, such as Buffaloes and Crocodiles, from 79th Armoured Division were used for the difficult conditions but most of the fighting had to be done by infantry on their feet. Artillery and air bombardment were used to the full and again there were occasions when the air bombardment had the opposite effect to that which was intended. For example, 1,384 tons of high explosive was dropped on Cleeve and the resulting chaos considerably delayed XXX Corps. This time it was not the army's fault because Horrocks had asked for support by incendiary attack. The Rhineland battle, like El Alamein, the Italian fighting, and the battles in Normandy, showed once more that there was no easy solution to the problem of attacking an enemy in prepared positions where there is no open flank. These were all battles in which a technique not dissimilar from that of August 1918 was used.

The Rhine crossing saw all three services engaged once more in a combined operation. The Rhine flotilla took part and a Royal Marine Commando Brigade captured Wesel. Airborne forces were used for the first time since Arnhem and this time they landed close in; they were in fact within the range of artillery on the west bank. However, the defeat inflicted on the enemy in the Rhineland battle together with considerable success by the Americans further south, notably the dashing capture of Remagen bridge, meant that the crossing called more for organization and technical skill than hard fighting. During the Rhineland battle Second Army had carried out its task faultlessly. All the difficulties of getting forward the bridging and the enormous quantity of stores close behind the fighting area had been overcome despite the dangers of congestion at the Meuse crossings. Staff work and traffic control in the British Army had changed for the better since Loos.

Chapter 13

THE WAR AGAINST JAPAN

The war against Japan is especially interesting to the British because it sees their emergence from complete and humiliating defeat in Malaya to absolute victory in Burma. The campaign was fought almost entirely in undeveloped countries; indeed some operations were carried out in jungle to a depth not previously thought possible for organized troops. Yet it was in undeveloped countries that generations of British and Indian soldiers had learnt the lessons of warfare: with Wolseley in Burma, Ashanti, Zululand, Egypt, and the Sudan; with Roberts in Afghanistan; a great army in the veldt of South Africa; and countless small operations on the North-West Frontier of India. Here was none of the highly developed warfare in the technique of which the Germans were so proficient, but a jungle-bashing, hill-climbing war which might be expected exactly to suit the British and particularly the Indian Army. But it was a modern war too; no land campaign has ever brought out better the strength and limitations of air power, never has greater skill and ingenuity been necessary to overcome the administrative difficulties, and never has an enemy been brought to battle that took the injunction to fight to the death more literally than did the Japanese. One small part of the war, the two Chindit expeditions, can be thought of as the continuance of the type of irregular operations with which Lawrence had captured the imagination of the British public during the Palestine campaign of the Great War.

Although the manner of the Japanese entry into the war in November 1941 was a surprise, the invasion of Malaya was not unexpected. The fact that it found the British ill prepared was in large measure due to their having taken a risk with their eyes open. They were fully committed in the war against Germany and Italy, they had in Malaya and Singapore sufficient land forces for its defence, they had formidable defences against seaward attack. Moreover the United States had a large navy in being, including powerful aircraft carriers, and would be unlikely to stand idly by if the Japanese entered the war against Britain. It was not to be expected, therefore, that the British would send to the Far East those aircraft and naval ships which were so urgently required elsewhere. On the other hand, it was realized that the defence of Malaya was primarily an air and sea problem. The army could fight to hold the frontier and the most likely landing places, but only the navy, and above all the air force, could prevent each defensive position being outflanked all down the length of the Malayan Peninsula. The Commander-in-Chief was an airman, Air Chief Marshal Sir Robert Brooke-Popham. In January 1941, almost a year before the Japanese attacked, the Chiefs of Staff in London agreed with the Commander-in-Chief's estimate that 582 aircraft were required for the defence of Malaya, but they also pointed out that even the 336 aircraft allowed for in their own previous estimate would not be available before the end of 1941.[1] At this time the priorities were for the air war against Germany and for operations in North Africa; few aircraft could be spared for a theatre the war might never reach.

Despite the fact that the necessary number of aircraft were not provided the Royal Air Force remained primarily responsible for defence. This fact and indeed the geographical and strategic requirements of defence meant that the security of our Allied airfields was of paramount importance. This problem of the security of airfields was not however approached in the logical way of siting the airfields in accordance with defence requirements as a whole, thus putting them where they could best be defended. Instead airfields were sited where they would give maximum range over the sea approaches to Singapore and across the Malayan frontier. The army was then required to defend them. The airfield at Alor Star, situated in the artificial salient on the frontier, and the airfields at Kota Bahru and Kuantan all imposed a dispersion on the army which Percival was not strong enough—and could hardly have been strong enough—to resist.

Although the principal threat was from the sea, the land threat had not been discounted. In 1938 General Dobbie, the

[1] Kirby, *The War against Japan* (HMSO, 1957), vol. I, p. 54.

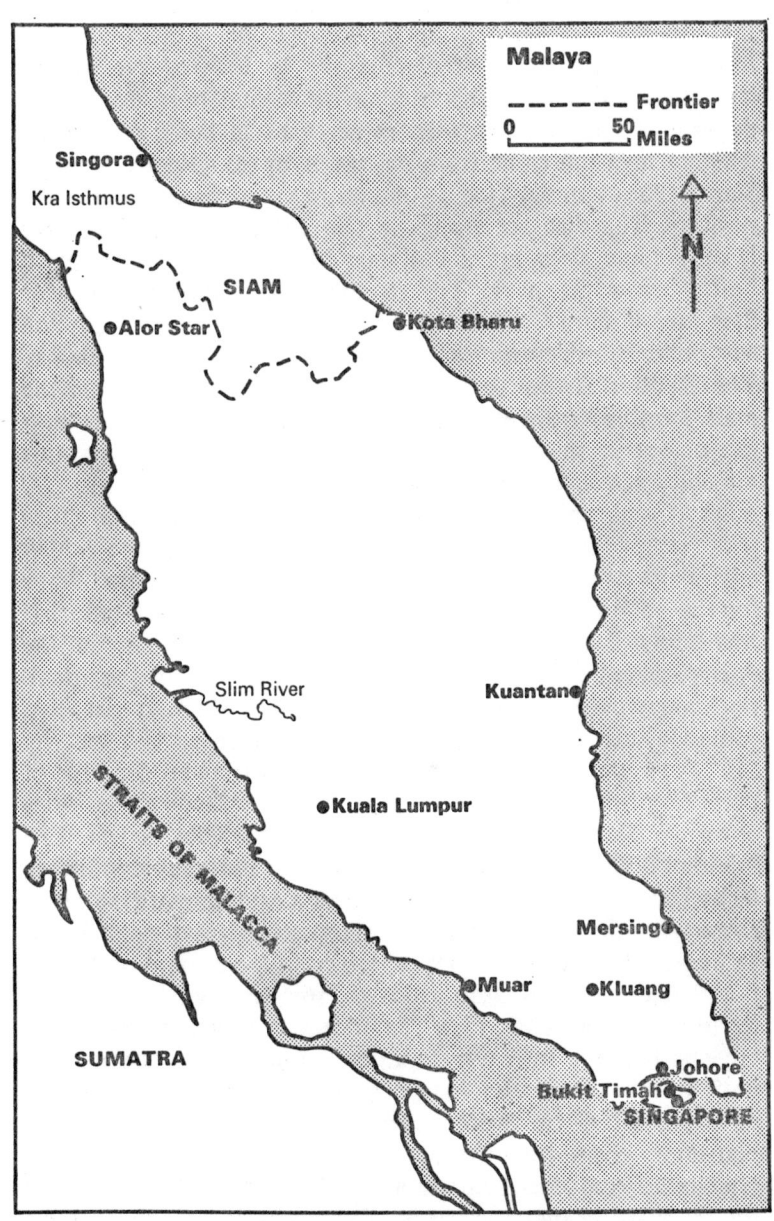

Map 12.

then commander, had pointed out that a natural adjunct to the seaward defences of Singapore was the construction of formidable defences on the mainland in Johore.[2] No money had been allotted for this and no work had been done. The defence of Malaya was seen rather as a task for air forces, whether invasion came from the sea or by an advance through Siam. The frontier itself offered an indifferent defensive position. The distance from coast to coast north of Alor Star is only some sixty miles but then Malaya widens out like a bottle below the neck. The frontier curls south round Alor Star and it flanks the coastal plain some 70 miles from the east coast before turning to reach it just north of Kota Bahru. The frontier is thus some 250 miles long and it leaves the main road and railway approach to Kota Bahru in Siam. With the co-operation of Siam, or by violating its territory, this difficulty could be overcome, since by advancing some 200 miles it was possible to take up a position only fifty miles wide across the Kra Isthmus. This position had the added advantage of covering the port and airfield of Signora. Plans for the advance were made under the code-name Matador. It was hoped that if she were threatened Siam would invite the Allies in, and that in any case sufficient warning of the attack on Siam would be received to enable the Allies to put Plan Matador into effect. The advance was to be carried out by III Corps under Heath. The corps was also given the task of preparing alternative but much less satisfactory positions covering Alor Star and the approaches to Kota Bahru. The inherent difficulty of giving the same force mutually conflicting tasks became apparent as the Japanese operations developed. Even when it was obvious that Siam and Malaya were about to be invaded there was opposition both from Siam and from the British political authorities to any violation of Siamese territory. Nevertheless Plan Matador was militarily so advantageous that the intention to carry it out was not abandoned until it was clearly no longer possible. Kept poised for the advance, III Corps were neither allowed to carry out their plan nor given time to occupy their alternative defensive position. From the beginning therefore the Allies had lost the initiative in the land battle. The overriding problem of the

[2] *Ibid.*, vol. I, p. 15.

The War Against Japan

military commander was to regain the initiative; the chief military interest of the campaign is to discover why he never looked like doing so.

The initiative is one of the most difficult of all elements in war to define. In reading the story of a campaign it is easy to see which side has the initiative, but not always easy to see why. In battle the ability to hold or to gain the initiative is the mark of the genius of the commander. Although it is well known that God is on the side of the big battalions the initiative by no means always goes to the larger force. In Malaya the Japanese land forces were never superior to those at Percival's disposal. The primary cause of the Allied loss of the initiative was that they lost command of the air, and by that very fact also of the sea. At the very moment when the Japanese were launching their invasion of Siam and North Malaya from the sea the Allies suffered a catastrophe by the loss of their only two capital ships, Prince of Wales and Repulse. They were lost in an action in which our air forces did not even attempt to participate. The Allies were at that moment, through no fault of the air forces in the area, quickly losing the air battle, but it is unthinkable that there should have been no attempt to protect the ships from air attack if both navy and air forces had been under the same Commander-in-Chief. But even if this catastrophe had been averted at this moment it is hard to see how much longer the Prince of Wales and Repulse could have operated within reach of Japanese air forces. The Japanese had a superiority of four to one in aircraft and their Zeros were superior to anything the Allies then had in the theatre; the air force was thus virtually destroyed in twenty-four hours. The loss of the Allies' two great ships and the destruction of their air force meant that the enemy could choose at will the beaches where he would attempt to land from the sea. Only determined resistance by the land forces on the beaches could prevent such landings; thus a number of defensive requirements accentuated the tendency towards dispersal. In the face of the Japanese tactics of outflanking Allied positions by every available track through the jungle and by a number of short hooks by sea, the British land forces showed themselves incapable of imposing their will on the enemy either by establishing an effective area of resistance or by launching the counter-attack which would destroy the main invading force. The bold tactics of the Japanese in Malaya laid their forces open to destruction and such destruction should have been the task of the reserves in the hand of the commander. The British forces which could have acted as such a reserve were engaged in watching other landing places in Southern Malaya and Singapore Island.

To return to the operations of III Corps, the position at Alor Star was soon lost, outflanked by the direct route along the eastern road and railway, where British delaying actions had not been well handled. At the same time the enemy effected a landing at Kota Bahru. The Indian Brigade there fought well and the situation might well have been held were it not for the withdrawal of the remainder of the corps which failed to hold the position north of Butterworth and the approaches to Penang Island. Up to this time it can be said that the losses of territory were no more than might have been expected after a surprise coup by the enemy and some delays and mistakes in first meeting it. The time had now come, however, when Percival had to gather together his forces and restore the situation. The essential requirement in fighting a campaign forward in Malaya was to gain time. The task was to hold Singapore and sufficient elbow room in Johore for long enough to allow reinforcements, particularly air forces equipped with modern fighters, to be landed. Despite the disasters that had befallen the Allies at sea and in the air the task demanded of the army was not beyond its powers. But this task could only be achieved if the army knew what it was doing; if it realized that delay imposed on the enemy did give time for the placing of forces for the decisive battle, and if withdrawals were made with the purpose of eventual participation in this battle. Percival had an almost impossible task. In May 1941, when he came to Malaya, he was subordinate to the Commander-in-Chief Far East. He therefore lacked the authority to insist on the co-operation of the civil government. By the time the fighting began the Commander-in-Chief had gone without replacement in Malaya. Thus when he ought to have been free to fight the battle Percival was saddled with all the responsibilities of participation in the Colonial Government. We see the absurd situation of Percival hurrying from a

The War Against Japan

Governor's Council to give operational decisions. Doubtless Percival could take refuge in some degree of decentralization but nothing could absolve him from responsibility of command of the battle, and nothing short of his intervention—and often not even that—could move the civil government to effective action in such matters as control of labour, intelligence, and civil defence measures.

Much as one can sympathize with Percival in the dilemma which faced him, it cannot be said, even from his own account of operations,[3] that he ever made up his mind what he wanted to do nor did he impress his will upon the battle. There is no evidence that he ever directed his commanders' minds towards any plan that might regain the initiative, or to any area in which a decisive battle might be fought. Instead the only plan in their minds was to hold on as long as possible. Such dates as were discussed for holding various lines always proved hopelessly over-optimistic because the enemy had outflanked them and there were no reserves to deal with the situation. The resulting losses in men and material, and inevitably of morale, frittered away III Corps. In the record Percival gives of his conversations with his subordinate commanders it is noticeable that the suggestions always come from them. It was never that he told them what to do but that they told him what they must do. The result of the losses and disorganization during the retreat was that Percival's only intact field formation, 8th Australian Division (consisting of two instead of the usual three brigades), was required to take up a defensive position instead of being available for a counter-attack role.

No criticism of Percival can be just without some examination of the courses which were open to him. It is the measure of his task that even with hindsight and in the comfortable role of armchair critic, it it not easy to decide what he could have done. Once the frontier battles were lost there seem to have been two courses open to him. Either to fight north of Kuala Lumpur in the so-called Slim River position or to retire further south about Muar and Kluang. The Slim River position would have retained a large part of Malaya, including Kuala Lumpur the capital, Malacca, and the narrow portion of the Malacca Straits. But there the peninsula is almost at its widest

[3] Lieut.-General A. E. Percival, *The War in Malaya* (Eyre & Spottiswoode).

and separate battles might have to have been fought on the east and west coasts. Moreover, the position would have been outflanked by an enemy landing at Kuantan or Mersing. The much shorter Muar position, which could incorporate the possible landing areas at Mersing, offered a much better position. Even so it was close enough to Singapore to be regarded as the last possible position and the decision to retire so far without serious resistance would have been a bold one, bringing down on Percival's head much political criticism including that of Churchill. Nevertheless, it seems likely that the course of operations would have been very different if, say in mid-December, it had been made clear that the main battle would be fought on the line Muar - Kluang - Mersing. Operations by III Corps north of this line would then have been delaying actions; the running fight and the uncertainty and disorganization would have been avoided. 8th Australian Division would have been available to Percival as his reserve and main striking force. Knowledge of what happened strengthens the idea that they would have been admirable in this role. It is true that their commanders at all levels were inexperienced but judgment of what happened must be tempered by knowledge of the situation when they eventually went into action in January. The battle for the mainland had by that time been lost. The division was not then a sufficiently large force to retrieve the situation. Units fought magnificently, but there were mistakes at command and staff level which were almost inevitable in such a situation. Later, in the fight for the island, the Australians bore the brunt of the main Japanese attack. Again the units fought magnificently and again there were mistakes at command level. But from a reading of the Official History[4] it seems evident that given a definite role, preferably a counter-attack role, in an organized defensive position they would have been a match for any Japanese division.

No judgment of the fighting would be complete without some discussion of the opposing armies. It must at once be confessed that the British (i.e. British, Indian and Australian forces) had not given sufficient attention to jungle warfare and the problems it raised in attack and defence. But it would be a

[4]Kirby, *op. cit.*, vol. I, ch. 23.

The War Against Japan

mistake to imagine that the Japanese were adept at jungle warfare in the way that the Allied troops became in subsequent campaigns. The Japanese simply used jungle tracks that were known to exist and by short detours outflanked the Allied troops in defensive positions and ambushed them on the move. They also used the sea for short auxiliary outflanking movements to an extent which the Allies never found possible—or at any rate expedient—even in subsequent campaigns where they had almost complete air and sea superiority. These bold and elementary tactics mastered like a drill were accompanied by a tenacity and a willingness to die rather than to fail which when first encountered produced the myth of the unbeatable enemy. This myth undoubtedly played a large part in the Japanese victory.

There were many fine commanders at all levels and some first-class British and Indian units in Malaya at the end of 1941 but it is true that the Allied army was raw and less well trained than it ought to have been. Both British and Indian units had lost considerable numbers of officers and non-commissioned officers to act as reinforcements for the Middle East and to raise new units. In the Indian Army, which had perhaps been too ambitious in its expansion programme, this was particularly disadvantageous. At all times the Indian Army has depended on British officers who know their men, their habits and characteristics, and above all their language. It was this corps of officers, backed by experienced Indian officers and non-commissioned officers, that made the Indian Army the magnificent fighting machine it showed itself in the Middle East, Italy, and subsequently in Burma. The Australians were, as always, magnificent fighting material but, as has been said, the leadership was inexperienced. Most grievous mistake of all, a large contingent of Australian reinforcements had been sent to Malaya in time to fight. Instead of sending trained reinforcements from the camps in the Middle East, untrained men had been sent from Australia, quickly formed into improvised units, and put into battle.

A careful reading of the Official History discloses many mistakes and disasters but in the fighting on the mainland there seems to be only one place which points to the definite failure of a commander to fight to carry out the task allotted to him. In this the unauthorized withdrawal of a brigade astride the railway in South Johore led to the destruction of another Indian Brigade the safety of whose withdrawal depended upon them.[5] In the fighting on the island there were again mistakes without which the Japanese would not so easily have gained their footing on the north-western coast. But it was not there that the battle for Singapore was lost but later in the unauthorized withdrawal of an Australian Brigade astride the Bukit Timah road. This let the Japanese thrust between the 12th Indian Division still holding out in the east and the Australian and Indian units holding the Jurong line. It may be said, however, that under the circumstances the loss of Singapore was inevitable once the Allies had withdrawn from Malaya. The island itself was not defensible from the north unless a substantial part of Johore was held also.

In conclusion, it must be said that the army in Malaya was less well trained than it should have been but that, even allowing for the defeat at sea and in the air, it should have been a sufficient instrument to hold the southern half of Johore and Singapore sufficiently long to enable air and sea superiority to be regained in the area. The reason it did not prove sufficient is that the Allies failed to concentrate on the essential aim of defeating the Japanese army. We dispersed our forces in an attempt to defend everywhere, thus giving one more proof of the military aphorism, ' He who defends everything defends nothing '.

When Brooke-Popham had been relieved as Commander-in-Chief it had been intended to replace him by Lieutenant-General Sir Henry Pownall. Before Pownall arrived, however, it had been decided to set up an Allied command for the whole South-Eastern Asia area. Accordingly on 3 January 1942 Wavell became Supreme Allied Commander American, British, Dutch Area (ABDA), with Pownall as his Chief of Staff. The area included Burma, Malaya, the Philippines, and the Dutch East Indies. It was too late for Wavell to exert any real influence on the situation in Malaya, or indeed in the area as a whole, and by the end of February the command was dissolved. The defence of Burma remained the responsibility

[5] *Ibid.*, vol. I, pp. 336-338.

of Wavell who again took up his duties as Commander-in-Chief India.

The Japanese began their attack on Burma before the fall of Singapore and the campaign began in earnest with a disaster on the Sittang River so that Rangoon was soon lost. Command was now in the hands of generals who were later to show their outstanding qualities. Alexander was in command of the army, and soon after the fall of Rangoon Slim took over the Burma Corps, Alexander's remaining forces being a Chinese contingent under the American Lieutenant-General Stilwell. There is no question but that in this phase the Japanese proved themselves better trained and more skilful than our forces or the Chinese. But in the hard fighting and the long retreat both British and Indian units were learning. They were proving that they could yet be a match for the Japanese and they gained valuable time for the complete upheaval which was necessary in India to prepare for the defence of her north-eastern frontier.

The only likely threat to India had always been considered as coming from the north-west. The wild undeveloped North Eastern Frontier had been thought to give its own defence. Land communications with Burma did not exist but entry into Burma was through Rangoon and a few other small ports. From Rangoon communications ran northward along the grain of the country and through the valleys of the great rivers. From the railhead at Lashio the Burma road ran through Wanting to China. The climate of Burma too, was considered a formidable bar to large-scale operations. The monsoon brought torrential rain to the country from about mid-May until the end of September. Movement off the roads became impossible, and streams and rivers became raging torrents. Malaria, scrub typhus and most tropical diseases were rife.

As the monsoon broke in May 1942 our forces had crossed the River Chindwin and the Japanese were in possession of virtually the whole of Burma. Even allowing for their scant administrative needs it was clear that the Japanese were at the end of their communications. But they were in possession of the port of Akyab with its airfields and they still had a degree of air superiority which gave them control of the Bay of Bengal. Although it appears from post-war research that the Japanese had no intention of invading India we did not know this. We saw the threat to East Bengal, possibly to Calcutta. Even if a major invasion of India was not attempted there was a very real fear that limited operations would be accompanied by an attempt to organize dissident elements in India to overthrow the British Raj.

By the end of the monsoon neither side was ready for the offensive. Japanese losses in the Pacific, including half their strength in aircraft carriers, at the battles of Coral Sea and Midway in May and June, had turned the balance of sea power against them. The British had ambitious plans for an amphibious attack on Rangoon combined with operations from Assam to link up with a Chinese offensive into Burma. But the loss of the summer battles in North Africa meant that none of the promised reinforcements for the Far East could be provided. Moreover, the differences between the British and American view of the war against Japan[6] had already begun to appear. The Americans saw India only as a base from which communications with China could be established by air and by the construction of a new road linking the Assam railhead at Ledo with the Burma road north of Lashio. Thus from all other points of view the Burma theatre was not only lower in priority than operations against Germany but also bottom of the priorities for the war against Japan. The problem faced first Wavell then Mountbatten and it is in this perspective that the task to be given to Slim in Burma must be seen.

In the absence of resources for major operations in 1943 the Allied efforts were confined to an attempt to capture Akyab and a small scale operation by light guerrilla type forces beyond the Chindwin. Neither achieved important strategic results but both had a far-reaching effect on the outlook and organization of our army in Burma. The operation against Akyab had been intended as a combined land, sea and air operation but the amphibious craft required could not be made available. Slim who was now commanding the XV Corps was free for the operation but General Irwin at Eastern Army Headquarters at Barrackpore, just outside Calcutta, preferred to exercise command direct. The attempt to command from what was to all intents and purposes a peace-time headquarters was doomed to failure, and although Wavell

[6] See Ch. 10.

from Delhi and Irwin made frequent visits to the front the battle lacked a single controlling hand.

The approach to Akyab by the narrow Mayu Peninsula, with its precipitous jungle covered ridge along its length and its intersection by many steep watercourses, presented a difficult tactical problem. The unwarranted assumption that the Japanese could neither cross the ridge nor outflank the peninsula resulted first in a dispersal of forces then in an attack on a narrow front using tanks in penny packets. Not only did the attack, pressed several times along the same line, fail, but the Japanese counter-attack brought them again to the frontier and the approaches to Chittagong, the starting point of the attack.

The sense of failure and the loss in morale due to defeat for which there was no sound military reason was to some extent offset by Wingate's first expedition in February to April 1942. This expedition had been originally planned in concert with a Chinese offensive from Yunnan. When it became clear that this offensive would not take place Wavell authorized Wingate to go ahead with an isolated raid; its object was to cut Japanese communications on the Mandalay-Myitkyina railway and, if Wingate thought it feasible, then to cross the Irrawaddy to cut the railway to Lashio. Wavell had known Wingate in the troubled times in Palestine before the war, and later when Wingate had led a force of partisans in the reconquest of Abyssinia. Wingate was a man with a mission. He believed that he could lead men to defeat the Japanese at their own game. He not only believed in himself but he had the power to make his men believe in him. With others than his chosen followers he was obstinate and difficult. At this time Wavell was almost the only person in authority who believed in him. After the expedition he was to get the ear of Churchill and Roosevelt. It is however significant that Wingate is almost the only man for whom Slim has a hard word to say in his fascinating account of the campaign.[7] On the other hand fighting men as diverse as Bernard Fergusson and Mike Calvert and the hard-bitten rank and file of their columns had implicit faith in him.

The Chindits, as they were called, crossed the River Chind-

[7] Slim, *Defeat into Victory* (Cassell).

win in five columns on foot and, entirely maintained by air, succeeded in destroying a number of bridges and cuttings on the Myitkyina railway. Wingate then decided that he would attempt his second and permissive task and ordered his columns across the Irrawaddy. By this time the Japanese reaction was strong and the Chindits were faced with the problem of operating in open country for which they were not equipped. Wingate then ordered the columns to split up into small parties and make their way back as best they could across the Chindwin into Assam. In blunt terms of military achievement the expedition had failed. The damage to the railway was quickly repaired, no major Japanese formation was long diverted from its purpose, and about a thousand men (a third of the Chindits) failed to return. The cost in air effort too was considerable but two things had been achieved; the Allies had shown that they could survive behind the enemy lines and that they could fight and defeat the Japanese. The technique of jungle operations and of air supply which the Allies had evolved was to stand them in good stead for the remainder of the war and beyond. Skilfully handled, the British press made great use of the epic stories of the expedition and did much both at home and in America to offset the failure in Arakan. There was to be a later dividend too from the expedition since it made the Japanese doubt the security of their Chindwin line.

After these operations the Allies set about serious consideration of their command structure. Mention has already been made of the difference in strategic outlook between the Americans and the British, but the Americans were prepared to accept British command of Allied forces in South East Asia provided it was understood that their forces were in India with the purpose of establishing lines of communication with China and for the support of the Chinese armies. They did not mind what the British did with their residual forces provided they could facilitate Stilwell's operations in North Burma and the construction of the road to link Ledo to the Burma road. Accordingly in August Admiral Mountbatten was appointed Supreme Allied Commander South East Asia with General Giffard under him as commander 11th Army Group which was to consist of Fourteenth Army under Slim and Northern Combat Area Command, principally Chinese troops, under Stilwell. Stilwell was also Deputy Commander to Mountbatten, and also owed allegiance direct to Washington as Commander of all American troops in India, Burma, and China. At this time it must be remembered the United States Air Force had not been created; the air forces operating in Burma and China were army air forces and so directly under Stilwell's command. To add to his responsibilities Stilwell was Chief of Staff to Chiang Kai Shek. To make things even more complicated Stilwell refused to serve under Giffard, but he did agree to serve under Slim's operational direction. The air forces were to play so vital a part in the campaign that it is worth trying to see, in simple terms, how command was exercised. The air forces that concern us, that is those excluding the maritime air forces and those allotted specifically to the support of Chiang Kai Shek, were organized as Eastern Air Command under the American General Stratemeyer. This was an integrated Allied force with the British providing the higher proportion of tactical aircraft while the Americans provided most strategic, transport, and cargo aircraft. Both Air Marshal Baldwin, commanding 3 Tactical Air Force and the American Brigadier-General Old of Troop Carrier Command worked in headquarters adjoining Slim's and the three were to all intents and purposes a single headquarters.

From the moment when he took command of Fourteenth Army Slim saw that he had it within his capacity to forge and use the weapon that would, in his own phrase, turn defeat into victory. Despite the many plans that were hatched for amphibious operations, plans that were always eased out because of the low priority of the South-Eastern Asia theatre, and despite the faith the Americans put in Chinese offensives which never materialized, Slim saw that he would have to defeat the Japanese in Assam and Central Burma. He saw the three facets of the task which lay before him. First he had to have an army in which the morale, the fitness, and the training were at the highest pitch. Secondly he needed an administrative team which, under his leadership, could surmount every difficulty of terrain, of lack of communications, and of climate. This team by the use of every possible expedient and improvisation could see that his army was supplied and maintained where he wanted it to fight. Lastly, a facet dependent on the other two but dependent also on his own spark of military genius, he had to devise the tactical means to wrest the initiative from the Japanese and finally to defeat that hitherto irresistible enemy.

To take the last point first, Slim saw that the Japanese tactics always followed the same pattern. They relied on outflanking movement to cut Allied communications and so force a withdrawal. Every withdrawal meant loss of equipment, loss of men, and loss of morale, so that when the Allies did fight they did so at a disadvantage. But the Japanese outflanking movements obviously laid themselves under the danger of having their attacking forces isolated and destroyed. Slim realized that the British ability to stand their ground and to fight it out despite an enemy astride their communications was only part of the solution. To ask men to do this was only reasonable and only likely to be effective if the commander could ensure that they would be supplied with the necessities to fight long enough and that action would be taken to restore the situation and so to make their resistance worth while. The first requisite could be ensured by supply from the air; this was now becoming a real possibility owing to the change in the air situation. By the end of 1943 the Allies had a considerable degree of air superiority, and their technique of air supply had been well developed, particularly during the first Chindit expedition. The winning of the battle depended on the way the commander grouped his forces and on the intelligence which enabled him to deduce what the enemy could and would do. Shortage of forces to provide the reserves in the commander's hand was no longer one of our major problems; the difficulty was to have them in the right place at the right time and properly trained and equipped to fight. Here again good intelligence and good administration were necessary. So Slim's first task was to persuade his men that they could hold out and that they would only be asked to do so when results would make it worth while. His second task was to produce the fighting machine capable of destroying the enemy's attacking forces. If he could achieve that in a major battle the initiative would pass to him and he could turn his efforts to the complete destruction of the enemy.

Slim saw the problem of morale, health, and training as one

interdependent problem. The feeling of the British that they belonged to a 'Forgotten Army', the lack of understanding of the Indian of why the war was their concern, the effect of discomfort and disease in the rain-sodden jungle, and the memory of an endless series of defeats could not be cured by pep talks and home comforts. The three-sided problem of morale—spiritual, intellectual, and material—could only be solved by leadership. This leadership he gave and inspired himself and it passed down through all ranks of his army, bringing out the proper relationship between commanded and commander. Results were achieved not by promises of comfort and ease but by passing on the belief that it was a task within their capacity. A task requiring effort and involving hardship, but hardship which would be demanded only when essential to achieve results. Discipline and military skill were essential elements in the cure; it was easy to see the necessity for the second, but the first is not so often understood by those with no experience of leading men in time of stress. The discipline was human and inspired but none the less real, from the routine of dress and the use of mepacrine to defeat malaria to the requirement of obedience where to obey meant discomfort, danger, and sometimes death.

Morale for the fighting unit is always easier than for the mass of administrative units that populate the rear areas. Slim saw that the way to get results was to play on the human desire of every man to feel himself and his work important. In this war there was no lack of commanders who showed themselves and knew the conditions in the forefront of the battle, but it was necessary too to let the telephone operator, the clerk, the supply packer and a host of others know that their work mattered too, and that it was appreciated. Slim missed no opportunity of driving this home; in his talks to soldiers, many of them little educated Indians, he often used the simile of the clock. The army commander was the mainspring and the officers and men the wheels and cogs. Some had more important parts to play than others, but if any part failed the clock stopped.

In no branch of the art of war did Slim show his skill more than in his understanding of the relationship between tactics and administration. The commander who makes his plan regardless of the administrative factors and leaves them to his staff to arrange as best they can, is certain to run into disaster. But the commander who is bound by administration will never do anything. Slim showed his mastery by realizing what he wanted to achieve, what forces he wanted and where, and how that fitted in with what, on the face of it, was administratively possible. By thinking always one or more steps ahead, and by taking into his confidence early his Chief Administrative Officer, Major-General Snelling and his Chief Engineer Brigadier Halstead, Slim was able to balance his account. What were the best risks to take, reduction of forces or administrative shortages? How could administrative potential be improved, by cuts in scale, by airfield construction, by road improvements, by a different phasing of the operation? Whatever the solution self-help, invention, and the use of every possible device and improvisation were added for good measure. As an example of the kind of man he was Slim insisted that whenever the forward troops had to be put on a reduced scale of rations he and his staff suffered the same reductions, just to remind all concerned. The problems and the priorities differed in each battle, but we shall see examples of this same process of solution as we discuss Second Arakan, the Imphal-Kohima battle, the advance to the Chindwin, the master stroke of Meiktila, and finally the pursuit to Rangoon.

Slim was almost entirely dependent upon India for the maintenance and reinforcement of his army. Giffard the Army Group Commander provided an excellent bridge between Slim and India, where Auchinleck, now Commander-in-Chief, was determined everything should be done to overcome the deficiencies of India as a base. Giffard in no way interfered in the tactical conduct of the campaign but while he was in command he was always in Slim's mind and he saw that the forces that Slim wanted were provided and moved to the place where they were required.

In his plans for the winter 1943-44 Slim had one paramount care. He could not risk a reverse or a major withdrawal. He had promised his army a victory. Fourteenth Army was stretched over two widely separated fronts, Arakan and Assam. In Assam, the main front, he did not at once wish to take the offensive because he would inevitably become in-

The War Against Japan

volved in difficult country where there was little to do except drive straight against a strongly placed and fanatically determined enemy. He preferred, if possible, to fight on the Imphal Plain where he could use the strong air support now available to him and his slight superiority in armour. In Arakan Akyab still offered a tempting prize, although with the changed air situation it was, for the moment, less important than it had been. Thanks to the fears which the first Wingate expedition had inspired about the security of the Chindwin line the Japanese showed signs of playing into Slim's hands. It was clear that they were preparing for an attack into Assam. Slim decided therefore to prepare for the main battle on ground of his own choosing about Imphal, and in the meantime to engage in a limited advance in Arakan. Here he intended to make secure the approaches to Bengal by securing the road from Maungdaw to Buthidaung, the only lateral road in the Mayu Peninsula. Only if opportunity offered would the advance be continued to Akyab.

The Japanese had by this time decided to make sure of their hold on Burma by capturing the Imphal Plain, the only jumping off position for a major British offensive. As a preliminary to their main attack the Japanese intended to attack in Arakan with the purpose of drawing off Slim's reserves from the Assam front. Slim, however, was the first to attack and in January 1944 XV Corps advanced to secure the lateral road. The Japanese thereupon put their Arakan plan into operation. By their usual tactics they were soon astride our communications; they completely isolated 7th (Indian) Division and took the other division, 5th (Indian), in flank. This was not unexpected and this time the British did not withdraw. Now was the moment for Slim to put into action all he had thought about and all he had trained and planned for. Preparations had already been made to supply both 5th and 7th (Indian) Divisions by air, while Slim had 26th (Indian) Division in reserve and 36th Division ready to move up from the Calcutta area. By the first week in March the Japanese attack had been held and such of the attackers that had not been destroyed had been forced to withdraw. By the 15th XV Corps had resumed the offensive and captured the lateral road. The victory in Arakan was the first stage in Slim's fight to regain

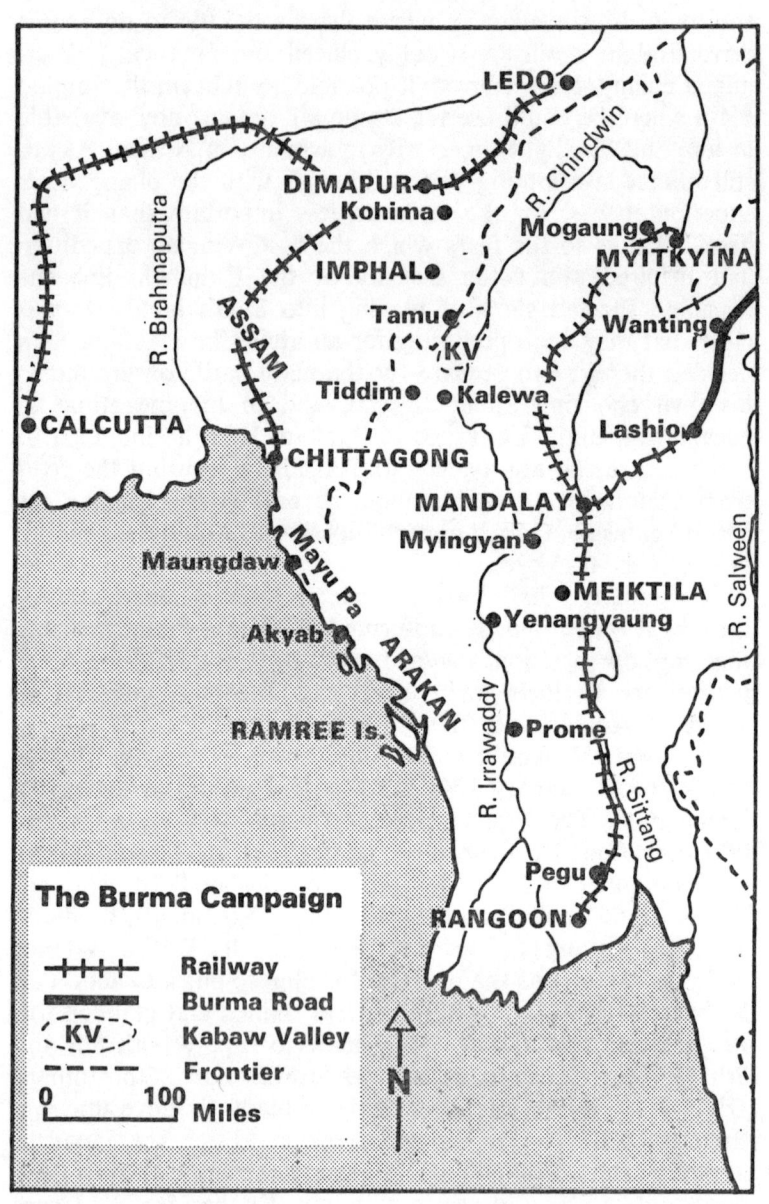

Map 13.

The War Against Japan 285

the initiative. It was of overwhelming value to morale; the old Japanese tactics had not succeeded this time, it was they who had withdrawn exhausted. Confidence in their leaders and in themselves was established in Fourteenth Army.

The offensive in Arakan was not continued because of even more important events on other fronts. Stilwell's advance from Ledo towards Myitkyina had begun and the new road was being built behind him. In support of Stilwell the first two of six Chindit Brigades under Wingate had been flown to areas deep behind the enemy lines and within striking distance of the Mandalay-Myitkyina railway; a third brigade was marching from Ledo towards Indaw. But it was on the Imphal front that the main battle had to be fought. The expected Japanese attack began on 15th March, two divisions (33rd and 15th) moved to outflank Imphal from west and north and a third (31st) to capture Kohima, our administrative base for the Imphal Plain. Scoones's IV Corps also consisted of three divisions, 17th (Indian) and 20th (Indian) watching the approaches from the Chindwin and 23rd (Indian) in reserve near Imphal. It had been Slim's intention to withdraw to the Plain before the Japanese attacked. There had been so many withdrawals before the Japanese that all commanders were loth to withdraw until it was quite certain the Japanese were going to attack. Slim therefore left the operational decision to Scoones, the man on the spot. By the time the Japanese had crossed the Chindwin our withdrawal had not begun, and, although 20th Division about Tamu got away intact, 17th Division, which two years before had borne the brunt of the withdrawal through Burma (under its same commander, Cowan) was slow to get away. They were outflanked on the Tiddim road and fought their way back magnificently, but by the time the Corps was established on the Imphal Plain 23rd Division, Scoones's only reserve, had been heavily committed.

In this situation Slim again showed his generalship and his forethought. A commander with no reserve can do nothing except make encouraging noises. Now that the situation in Arakan had been established Slim pulled out 5th Division and taking advantage of his earlier preparations moved it by air to Imphal. This was the first time an operational move by air of an ordinary division had ever taken place; everything went without a hitch but before the division was complete at Imphal a new danger had arisen, fears for the defence of Kohima. Slim had estimated that not more than a brigade group could move and be maintained over the difficult tracks to Kohima whereas the Japanese succeeded in using a whole division. Kohima was little more than forty miles from Dimapur, the railhead. For the defence of these vital administrative areas there was only one irregular brigade and the administrative troops working in the depots. This is where Giffard came in and where all the discussions and plans for eventualities paid their dividend. Orders were issued for the move up of a new corps headquarters, XXXIII Corps under Stopford, to command in the Kohima area and for 2nd Division to move from India to Dimapur and thence to Kohima. While this move was going on Slim appointed General Ranking to organize the Kohima garrison for its defence and moved the last of the brigades of 5th Division to Kohima instead of Imphal. One of the Chindit brigades was also taken for the Kohima battle. When Slim had moved 5th Division from Arakan he had made preparations for 7th Division to follow, and they also, less one of their brigades to replace the borrowed brigade from Imphal, went to the Kohima battle.

Before the battle was won Imphal had been completely cut off from Kohima and Kohima itself surrounded. Undoubtedly Sato, commanding the Japanese 31st Division, lost an opportunity by concentrating on the capture of Kohima and making no attempt to go for Dimapur. By grim fighting, supported and entirely supplied by air, IV Corps held out at Imphal and gradually gained the ascendancy. Then as XXXIII Corps increased in strength Kohima was relieved, and their thrust south combined with the break-out northward from Imphal. By the first week of June the crisis of the battle was over and Slim was thinking of regrouping his forces for the destruction of the Japanese.

While the Imphal battle was going on Stilwell was making good progress and he captured Myitkyina airfield on 17 May; but it was two and a half months before he captured Myitkyina itself. The part that the Chindits played in support of these operations is difficult to assess except that it is certain that their capture of Mogaung prevented the Japanese retaking Myitkyina airfield. Stilwell himself thought they could have done more, but as Slim said to him it was a little hard to reproach a few hundred men for failing to do what Stilwell's own 30,000 men had failed to do. The reason Stilwell was dissatisfied with the Chindits was that he did not understand or at any rate did not accept their role and methods. It is indeed difficult to say what this role was and should have been. Long Range Penetration Groups were the child of Wingate's mind; he had originated an excellent means of making raids deep in enemy territory, but by capturing the imagination of Churchill and others he had now built up a large private army and had made himself into a dictatorial personage that was difficult to fit into any regular military picture. This might have been no bad thing; there is usually a place in war for the unorthodox and always a place for inspired leadership. But two things went wrong; first Wingate himself departed from his own brilliant conception, and then early in the operation Wingate himself was killed. The basic idea was that Groups could be flown into strongholds behind the enemy lines. Stronghold is a name which gives a false idea because they were in fact areas which could be supplied by air but which could not be approached by heavily equipped enemy columns. The lightly equipped Chindits with their splendid training and endurance could thus deal with any enemy that could reach them, could raid and harry the enemy on their lines of communication, and at will could move to set up a new stronghold. They were not suitable to hold a defensive position against an organized and fully supported attack, nor were they equipped to attack a strongly held position. Before he was killed Wingate was beginning to preach the gospel of using his columns to hold a position which could draw major enemy attacks. After Wingate's death Stilwell was asking them to carry out attacks against organized positions. In operations like the defence of Hopin and the attack on Mogaung the Chindits wore themselves out in operations which were beyond their capacity. Moreover it was Wingate's intention that Chindits should not operate in the jungle for more than two months at a stretch and never in monsoon conditions. In the Second Expedition they were operating for three and a half months, half the time in the monsoon.

It would not be a slight to Lentaigne, Wingate's successor, or his magnificent brigade commanders to say that no one can tell what might have happened if Wingate had survived. On the other hand it is probable that the Chindits would have paid far better dividends if the large and costly expansion from two to six brigades had never taken place. Two brigades might have been more effective operating against enemy areas more closely concerned with the attack against Imphal or in closer concert with Stilwell's advance. It is relevant that Perowne's Brigade which was held back to fight on the flank of the Kohima battle did yeoman service in just such a role.

To return to the Imphal battle, it had been clear by the middle of May that the initiative was beginning to pass from the Japanese, but Mutaguchi, the Fifteenth Army Commander, was neither willing nor empowered to withdraw. As a result of this inflexibility and the skill with which Slim used his divisions, despite the onset of the monsoon, Fifteenth Army was practically destroyed west of the Chindwin. By the end of July the Japanese suffered more than 30,000 casualties (killed, wounded and missing) out of a strength of 53,000. Of those who remained with the units most were slightly wounded or suffering from malnutrition. British casualties were about 16,700.

As always, while the battle was still going on Slim was thinking out the next stage. The 11th East African Division was moved from reserve to Imphal and plans were made to continue the pursuit and capture crossings over the Chindwin. Slim's overriding consideration was to use the remainder of the monsoon period to put himself into a position from which the final blow could be delivered in the coming dry season. Administrative requirements would demand the early establishment of a bridgehead at Kalewa accessible by an all weather road from Imphal. IV Corps headquarters and several of the divisions required rest and training but, despite the deplorable conditions of the monsoon, XXXIII Corps continued the pursuit. The 11th East African Division moving from Tamu down the Kabaw valley and 5th Division were directed down the Tiddim road to eliminate all enemy west of the Chindwin and secure bridgeheads at Sittang and Kaiewa. These operations in conditions which formerly would have been considered impossible were entirely supplied by air, and by mid-November XXXIII Corps had completed its task.

Plans for 1945 discussed between Mountbatten and the Chiefs of Staff had given priority to amphibious operations to capture Rangoon, the tip of Sumatra and later Singapore and Malaya. These plans had come to nothing because the freeing of the necessary assault resources depended on the war with Germany being finished in 1944. Slim had early foreseen that he would be required to fight his next battle in Central Burma, and his monsoon advance had enabled the administrative preparations on which this battle would depend to be well under way by December.

Slim had now been relieved of responsibility for operations in Arakan, and XV Corps came direct under Leese who had replaced Giffard. The Japanese had also made changes; Kimura had replaced Kawabe in command of Burma Army and Katamura relieved Mutaguchi in Fifteenth Army. Slim was now within reach of the dry belt between the Chindwin and the Irrawaddy. He considered his best chance of a decisive victory lay in engaging the enemy in the Shwebo Plain which lies in the loop of the Chindwin and the Irrawaddy. Here his superiority in air and armour would come into play; he now had two tank brigades, one of them equipped with Sherman tanks against which the Japanese had no effective guns. Slim knew by experience the inflexibility of the Japanese and their reluctance to withdraw and he considered that they would fight west of the Irrawaddy to hold Mandalay. Slim's plan was to use IV Corps (now commanded by Messervy) to strike east from Sittang to the Irrawaddy and come down on Mandalay from the north. At the same time XXXIII Corps was to strike from Kalewa to cross the river below Mandalay. Each corps consisted of two divisions plus one brigade and one tank brigade, which was the maximum total force that could be maintained forward of the Chindwin. The northern thrust was led by 19th Indian Division, in action for the first time, and its rapid progress quickly persuaded Slim that the Japanese were not after all going to fight west of the Irrawaddy. He saw that if he persisted in his plan he stood to lose a whole campaigning season while the Japanese avoided a major engagement. The brilliance and speed with which Slim appreciated the true situation, changed his plan, and executed the new plan deserved the praise which Kimura gave it in describing it as the ' master stroke '.

Slim's new plan was that instead of a double thrust at Mandalay he would make a flank move to strike at Meiktila, the Fifteenth Army main base and communications centre. With a flank march of 250 miles on an unmetalled track from Kalewa necessary and the Irrawaddy still to cross, such a plan could only succeed if there was deception and surprise. The advance on Mandalay was therefore to be continued as a subsidiary operation. It would be undertaken by XXXIII Corps which would take command of 19th Division. The remainder of IV Corps, so far not committed, would undertake the flank march to Pakokku where assault crossings would be organized to allow the main strike to Meiktilla. In place of 19th Division Messervy was given 17th Division which was given an extemporary organization of two brigades with fully mechanized transport and one brigade lightly equipped for transport by air. The army reserve, 5th Division was similarly reorganized. South of Kalewa there was wireless silence and to complete the deception a false IV Corps Headquarters was left in its original location to handle signal traffic to and from 19th Division.

The move of IV Corps was not discovered and as they got to the required assembly areas for their assault crossing the deception was heightened by the activities of XXXIII Corps which crossed the Irrawaddy both above and below Mandalay, and by the move of 14th East African Brigade southwards towards the Yennenyaung oilfields. The Japanese were left with the firm conviction that Mandalay was the objective and that the moves down the river were feints and raids.

In order to make possible the operations by more than five divisions on and across the Irrawaddy reliance was placed first and foremost on air supply. Arakan now became strategically important again because only from Akyab and Ramree Island could airfields within economic range of Meiktila be found. XV Corps, including a Commando Brigade and a limited amount of assault craft, carried out the necessary offen-

The War Against Japan

sive. Akyab was evacuated by the enemy on 2 January and Ramree Island captured by 10th February. In addition to the move of bases for air supply into Arakan the lift was increased by flying more sorties per day than the normal limit and by the use of training squadrons. As the advance progressed both corps were required to select and make from their own resources forward landing grounds. Another measure to alleviate the administrative situation was the construction of home-made river craft in an improvised boat factory at Kalewa. These boats were used to carry forward stores and supplies for the establishment of an advanced base at Shwebo and later for a river-head base at Myingyan.

Leese had at his disposal a Parachute Brigade and this could have been made available to Slim. Slim decided not to use it because the air effort was better employed in the task of maintaining IV Corps.

The attack on Meiktila was made on 1 March by 17th Division and the Sherman Tank Brigade, soon followed by the air transported brigade of 5th Division. Meiktila was quickly captured and, as Slim said, he held the Japanese by the wrist so that their fingers along the Irrawaddy were powerless. He then proceeded systematically to destroy the Japanese Fifteenth Army between IV and XXXIII Corps. It was a grim and difficult battle because the enemy, as always, fought fanatically. Mandalay and Meiktila were both captured on 3 March but later the Japanese organized a counter-attack on Meiktila which involved fighting almost to the end of March before it was finally defeated.

Slim's aim from the first had been the destruction of the Japanese army in Burma. Fifteenth Army was now almost destroyed and Thirty Third Army from Northern Burma and Twenty Eight Army from Arakan, both already badly mauled in battle, were at his mercy if he moved quickly enough. Another essential consideration was that Rangoon must be captured before the monsoon began. Not only would this complete the reconquest of Burma, but it would also relieve an almost intolerable supply situation and would free forces for operations against Malaya. As the Meiktila battle went on, therefore, Slim was making his plans for the pursuit and as opportunity offered he took out his most mobile divisions, first 17th then 5th, under IV Corps to begin the drive for Rangoon. There were two possible lines of advance, the road and railway through Pegu and the road along the Irrawaddy through Prome. The first had two advantages; it gave greater scope for mobile and tank action, and, being further east, it gave the opportunity to cut off more Japanese from their communications into Siam. Slim decided that the mobile IV Corps with the Sherman tanks would use that route while XXXIII Corps, when it had completed mopping up, would use the Prome road and link up with XV Corps from Arakan. Slim saw however that the hard fighting at Meiktila had put him behind the clock. He therefore asked Leese to prepare for an amphibious attack on Rangoon in case if might be needed.

The chief problem of the pursuit was the maintenance of IV Corps, most distant from our airfields. The air-lift was only sufficient for one division and one brigade at a time so the solution was a system of leapfrogging. 17th Division captured an airfield and developed it for the further advance of 5th Division, and then vice versa. 19th Division followed to protect the lines of communication and to prevent a break-out to the River Salween.

2 May was the latest date the Navy thought an amphibious assault possible. On that day, as the first raindrops of the monsoon were falling, a Gurkha Parachute Battalion landed at Ferry Point, south of Rangoon. Before the amphibious brigade landed air reconnaissance disclosed that the enemy had left Rangoon. By this time IV Corps had reached Pegu, and XXXIII Corps were at Prome where XV Corps were soon to link up. Apart from the rounding up of the remaining Japanese the campaign was over.

The quality of Slim's generalship emerges clearly from the story of the battles in which he commanded. It shows above all his capacity to inspire and to train an army, to match the the tactical plan to the almost insurmountable administrative difficulties, to wrest the initiative from the enemy and then to mystify and mislead him and eventually to destroy him. But to appreciate Slim's achievement to the full it is necessary only to think back to the early paragraphs of this chapter. There we can see the feats of arms of which the Japanese were

The War Against Japan

capable and the aura of invincibility which they had created for themselves. It is true that the air situation in the two campaigns was very different, but air forces cannot by themselves win the land battle either for the attacker or the defender. And to Slim must go much of the credit for creating, in happy concert with his colleagues of the Royal and United States Army Air Forces, with the wholehearted support of Mountbatten and Giffard, and with the outstanding skill of the administrative staffs, the most effective system of air support ever given to an army in battle.

EPILOGUE

In little more than a generation from the time that Wolseley and Roberts quitted the active military scene, much that they had striven for had come about. Not all the lessons that the British might have drawn from the South African War were correctly applied but their experiences did lead to a growing conviction that soldiering was a serious business. There was in the army a nucleus of educated senior officers with the zeal and ability to apply themselves seriously to the organization and training of an army. In parliament and public life there was a similar element that realized that the organization of an army was a national and not only a military problem. When war came in 1914 the British put into the field an army which was for its size second to none, but which was pitifully small. The problem of the expansion of this army into a nation in arms was achieved with remarkable speed but with a woeful waste of human resources. The machinery of the government at war had many advantageous qualities but was marred in the early stages by too narrow a vision and later by a conflict between soldiers and statesmen in which there were faults on both sides. Judgment of Haig for his slowness in solving the tactical problem on the western front must be tempered by the knowledge that conditions there were more difficult than those faced by any commander before or since.

After the Great War there was an inclination among some officers to return to the care-free life interspersed with episodes of excitement and danger overseas; but those that mattered—and they were not to be found only among those in the highest places—were determined that the lessons of the Great War should be remembered and that those they led should not repeat the same mistakes. For the nation as a whole, however, the business of war was to revert to its amateur status with the hope that it would be all right on the day.

When the second war came the man to lead the nation and the right men to give military counsel and to command armies in the field soon came to the top. The machinery for the direction of war was apt for its purpose. There were days of grievous defeat and disaster before the British came through to the light and had allies to share their victories. In those dark days the men in authority bore an even heavier burden than those that were to come; then the exploits of Wavell and O'Connor stood out to light the future. In the long climb to victory so many played a part that it might be considered invidious to single out the few. But this cannot be so, for as Napoleon said, ' In war men are nothing; it is the man who is everything.' Alanbrooke was certainly the man who not only saw what had to be done but who understood the art of the possible. He was the architect of victory without whom Churchill's genius for leadership could scarcely have borne fruit, and he must take a place among the Great Captains. Montgomery and Slim, facing problems so completely different and yet each so exactly the right man in his place, would not suffer in comparison with Wellington on the field of battle.

There must always be some difference in outlook between one level of command and another and between the staff and the units. It is easy for the officer at divisional headquarters to feel that corps is out of touch with reality or that brigade is not getting on with the job, and for the man in a slit trench to feel that none of them care about him. Yet in the second war there was surprisingly little of this. Throughout the British Army commanders, staff, and units worked efficiently and happily together; commanders were known and trusted by those who worked under them and it was accepted by all that the duty of the staff was to serve. There was a return to the tradition of Wellington's heyday—despite the enormous difference in the size and dispersion of armies. Never since those days had relations between commanders, staff, and fighting troops been so good.

The British Army has the name for being the most conservative of institutions, yet its leaders proved how much they had appreciated the social and industrial changes which had taken place between the two wars. The use of man-power in the second war was in marked contrast to the first. There was none of the waste of the first magnificent flow of volunteers by drafting them to serve in the lowest ranks. Those who showed the attributes of leadership were quickly trained as officers and among these those whose profession or calling fitted them for executive or administrative work were trained for and employed on the staff—and many of the barristers, solicitors, schoolmasters, university professors and others made admirable staff officers. It goes without saying also that the engineering and industrial world was used both for our technical units and for the invention and production of weapons and equipment.

This book will have failed in its purpose if it gives the impression that the military art is something that can be learnt by rules or even by copying the movements and the stratagems of past masters. War is a two-sided struggle and usually a struggle in the dark. Certainly it is a battle of wills in which the commander cannot know what his enemy will do or what he is thinking. The lessons in war that are ageless are those that come from a study of the mind and will of the commander. Not what he asked a unit to do at a particular time, nor even the brilliant plans which he thought out in the quiet before battle, but how he imposed his will upon the enemy and upon his own men, how he recognized the possible without rejecting the difficult. Above all it is necessary to recognize through the commander's mind the meaning of the initiative in battle and to see how one on the point of defeat might yet retain or regain that initiative. It is usually fairly easy to know what to do but as Shakespeare said: ' If to do were as easy as to know what were good to do, chapels had been churches and poor men's cottages princes' palaces.'

Appendix

LIST OF SOURCES QUOTED

Official Histories, Reports etc.

Aspinall-Oglander. Brigadier-General C. F. *Military Operations: Gallipoli.* Heinemann, Vol. I, 1929; vol. II, 1932.
Butler, J. R. M. (ed.) *Grand Strategy.* H.M.S.O., 1956-7.
    Vol. II by J. R. M. Butler.
    Vol. II Pt. I by J. M. A. Gwyer. Pt. II by J. R. M. Butler.
    Vol. V by John Ehrman.
    Vol. VI by John Ehrman.
Edmonds, Brigadier-General J. E. (ed.) *Military Operations France and Belgium.*
    1914 Vol. I by Brigadier-General J. E. Edmonds. 3rd ed. Macmillan, 1933.
    1915 Vol. II by Brigadier-General J. E. Edmonds. Macmillan, 1928.
    1916 Vol. I by Brigadier-General J. E. Edmonds. Macmillan, 1932.
    1916 Vol. II by Captain W. Miles. Macmillan, 1938.
    1917 Vol. I by Captain Cyril Falls. Macmillan, 1950.
    1917 Vol. II by Captain W. Miles. H.M.S.O., 1948.
    1917 Vol. III by Brigadier-General J. E. Edmonds. H.M.S.O., 1948.
    1918 Vol. I by Brigadier-General J. E. Edmonds. Macmillan, 1935.
    1918 Vol. III by Brigadier-General J. E. Edmonds. Macmillan. 1939.
    1918 Vol. IV by Brigadier-General J. E. Edmonds. H.M.S.O., 1947.
    (All notes in the text refer to the editor *i.e.* Edmonds, 1915, vol. II, etc.)
Elgin Commission on the War in South Africa. *Report, including Minutes of Evidence and Appendices.* Four vols. Cd. 1789-1792. H.M.S.O. 1903.
Ellis, Major L. F. *Victory in the West.* Vol. I. H.M.S.O., 1962.
Greenfield, K. R. (ed.). *Command Decisions.* Prepared by the Office of the Chief of Military History, Department of the Army (U.S.A.). Methuen, 1960.
Kirby, Major-General S. W. *The War against Japan.* Vol. I. H.M.S.O., 1957.
Playfair, Major-General I. S. O. *The Mediterranean and the Middle East.* Vols. I and III. H.M.S.O., 1954 and 1960.
Roskill, Captain S. W. *The War at Sea.* Vol. I. H.M.S.O., 1954.
War Office. *Notes on Certain Lessons of the Great War.* 1934.
15th Army Group General Instruction No 34 dated 12 Jan. 1944. H.M.S.O.
Allied Army in Italy Operation Order No 1 dated 5 May 1944. H.M.S.O.

Training Manuals and Instructions etc.

Combined Training Part I Field Service Regulations 1905. H.M.S.O.
Field Service Regulations 1909. H.M.S.O. Pt. I Operations.
Training and Manoeuvre Regulations 1909. H.M.S.O.
Platoon Training 1919. War Office.
Notes on Recent Fighting No 19. General Staff, Aug. 1918.
The Fire Plan in the Infantry Brigade Battle. War Office, 1934.

Biographies, Memoirs and Military Studies

Asquith, Rt. Hon. The Lord Oxford and Asquith. *Memories and Reflections.* Cassell, 1928.
Beaverbrook, Lord. *Men and Power.* Hutchinson, 1956.
Belfield, Eversley and Essame, Major-General H. *The Battle for Normandy.* Batsford, 1965.
Blake, Robert (ed.). *Private Papers of Douglas Haig.* Eyre & Spottiswoode, 1952.
Bryant, Arthur. *Turn of the Tide.* Collins, 1957.
— *Triumph in the West.* Collins, 1959.
Carver, Major-General Michael. *Tobruk.* Batsford, 1964.
— *El Alamein.* Batsford, 1962.
Churchill, Rt. Hon. Winston S. *The Second World War.* Cassell, Vol. II, 1949; vol. III, 1950; vol. IV, 1951.
Cole, Captain D. H. *Imperial Military Geography.* Sifton Praed. 1921.
Collier, Basil. *A Short History of the Second World War.* Collins, 1967.
de Guingand, Major-General Sir Francis. *Operation Victory.* Hodder & Stoughton, 1947.
Esher, Viscount. *Journal and Letters.* Vol. II. Nicholson & Watson, 1934.
Eisenhower, General of the Army Dwight D. *Crusade in Europe.* Heinemann, 1948.
Fuller, Major-General J. F. C. *Memoirs of an Unconventional Soldier.* Nicholson and Watson, 1931.
Goutard, Colonel A. *The Battle of France.* Muller, 1958.
Hamilton, General Sir Ian. *Gallipoli Diary.* Two vols. Arnold, 1920.
— *The Commander.* Edited Anthony Farrar-Hockley. Hollis & Carter, 1955.
Henderson, Colonel G. F. R. *The Science of War.* Longmans Green, 1908.
Jackson, Major-General W. G. F. *The Battle for Italy.* Batsford, 1967.
Liddell Hart, B. H. *Foch, Man of Orleans.* Faber, 1931.
— *The Framework of a Science of Infantry Tactics.* Rees, 1951.
— *The Defence of Britain.* Faber, 1939.
— *Rommel Papers.* Collins, 1953.
— *Memoirs.* Two vols. Cassell, 1965.
Lloyd George, Rt. Hon. David. *Memoirs.* Four vols. Nicholson & Watson, 1933.
Ludendorff, General E. (English trans.). *My War Memories 1914-18.* Hutchinson, 1920.
Luvaas, Jay. *The Education of an Army.* Cassell, 1965.
Magnus, Philip. *Kitchener.* Murray, 1958.
Macleod, Colonel Roderick and Kelly, Denis (eds.). *The Ironside Diaries.* Constable, 1962.
Maurice, Major-General Sir Frederick. *Life of Viscount Haldane of Clone.* Faber, 1937.
— *Life of General Lord Rawlinson of Trent.* Cassell, 1928.
— *British Strategy.* Constable, 1929.
Monash, General J. *The Australian Victories in France.* Sydney, 1936.
Montgomery of Alamein, Field-Marshal Viscount. *Memoirs.* Cassell, 1958.
Muller, Commandant. *Joffre et la Marne.* Paris, 1931.
Percival, Lieut.-General A. E. *The War in Malaya.* Eyre & Spottiswoode, 1949.
Rhodes James, Robert. *Gallipoli.* Batsford, 1966.
Robertson, Field-Marshal Sir William. *Soldiers and Statesmen.* Cassell, 1926.

Appendix

Slim, Field-Marshal Viscount. *Defeat into Victory*. Cassell, 1956.
Spears, Brigadier-General E. L. *Prelude to Victory*. Cape, 1939.
Swinton, Major-General Sir Ernest. *Eyewitness*. Hodder & Stoughton, 1932.
Tedder, Marshal of the Royal Air Force Lord. *With Prejudice*. Cassell, 1966.

Periodicals

Journal of Royal United Services Institution. May 1920.
History Today. January 1966.

Unpublished Papers

Wolseley Papers. Royal United Services Institution.
Maxse Papers. Imperial War Museum.
Chetwode Papers. Imperial War Museum.
Great War Papers. Royal Artillery Institution.
Correspondence Liddell Hart & Wavell. Private possession Sir Basil Liddell Hart.
Salisbury Papers, Hatfield MSS. Quoted in Kitchener, Magnus q.v.
Private Correspondence in the possession of the Author and acknowledged in notes to the text.

Privately Published

Marriott, Major-General Sir John. *War Memories*. 1960.

INDEX

'A.B.D.A.', 274
'Aberdeen', 232-3
Abyssinia, 194, 220, 277
Achi Baba, 149, 151, 153
Acroma, 229, 231, 234
Adam, General Sir Ronald, 70
Adam, El, 230-4
Administration and logistics, 31, 69, 73-74, 155, 207, 225, 227-8, 230, 247, 252-3, 257-8, 265, 279-82, 285-92
Admiralty, 72, 114, 196
Age and responsibility, 28, 30, 154
Agedabia, 227-8
Agheila, El, 225-6, 228-9, 243
Air,
    attack (close support), 111, 119, 144, 190, 227, 237, 239, 255-7, 264, 286, 293
    bomber offensive, 193
    forces and R.A.F., 175, 192-3, 227, 229, 237, 241, 243, 256, 266-7, 279, 293
    Ministry, 196
    photography, 92, 115
    reconnaissance, 45, 49, 82, 115, 119, 144, 242
    superiority, 177, 193, 205, 216, 228, 239, 247, 251, 256, 273, 275, 280, 289
    transport and supply, 207, 258, 278, 280, 285-93
Airborne troops, 260-1, 264
Airfields, 253, 257, 266, 275, 282, 290-2
Aire, R., 143
Aisne, battle of, 50-4, 80, 134, 137
Akyab, 206, 275-7, 282, 290-1
Alamein, El, 199, 201-2, 214, 235-6
    battle of, 239-42, 253, 264
Alam el Halfa, 237-8, 241
Alanbrooke, Field Marshal Viscount, 79, 170, 179, 192-3, 197-214 *passim*, 225, 236, 295
Alban Hills, 245, 248
Albert, 133, 189
Aldershot, 16, 20, 43
Aleppo, 75, 77
Alexander, Field Marshal Earl, 173, 179, 198, 209-10, 214, 237, 243-51, 275
Alexander, Major-General H. T., 170 f.n.
Alexandretta, 57, 64
Alexandria, 62, 64, 201, 235, 238
Algeria, 201-2

Allenby, Field Marshal Viscount, 47, 73, 75, 77, 100-1, 104, 116, 147, 162
Allied Command, 45, 71, 76, 134-5, 137, 201, 207
Allied War Council, Supreme, *see* Supreme Allied War Council
Allies, differences between, 70, 84-5, 207-9, 241, 244, 253-4, 263, 278
Alor Star, 266
Amateurs at war, 99, 225, 294
American Civil War, 25, 41, 156
Amiens, 47, 76, 133, 136-8
    battle of, 142-4, 174, 214, 264
Amphibious operations, 205-6, 243-4, 252, 270, 273, 276, 279, 292
Ammunition, 73, 83-4, 87-8, 90-1, 133-4, 137, 143
Andaman Islands, 204, 206
Anderson, Lieutenant-General Sir Hastings, 170
Antelat, 227
Anti-Aircraft Defence, 168, 178, 193, 217, 260, 262
Anti-Tank guns, 122, 168, 178-9, 213, 217-8, 220, 223, 237, 255, 289
Antwerp, 45, 54, 81, 188, 256, 259, 261-2
'Anvil' (Dragoon), 208-9, 249, 258
Anzac Corps, 61-2, 149-53
Anzac Cove, 62-3, 151, 153-4
Anzio, 244-9
Arakan, 278, 282-6, 289-92
Ardennes, 145, 259, 262-4
Argentan, 256
Argonne, 145
'Arme Blanche', *see* Cavalry
Armentières, 133
Armistice 1918, 76-7
Armitage, General Sir Clement, 170, 182
Armoured fighting vehicle, *see* Tanks
Armoured force, 169, 177, 180, 190, 215, 218-9, 238, 244
Army,
    Australian, 76, 140-4, 187, 195, 214-5, 235, 271-4
    British,
        Armies: First, 100, 109, 144; Second, 73, 96, 104-6, 109-12, 145, 255, 262, 264; Third, 100-2, 113-6, 120, 144-5; Fourth, 89, 91, 96, 138, 143, 144; Fifth, 73, 96, 105-6, 109-12, 134, 138, 140; Eighth, 219-52; Fourteenth, 278-93

301

Army, British (contd.)
  Brigades: 1st Cavalry, 48; 2nd Armoured, 228, 230; 4th Armoured, 220, 222, 225, 227, 230, 232; 7th Armoured, 220, 222-3, 225; 22nd Armoured, 216, 220, 222-3, 225, 227-8, 230, 232-3; 1st Guards, 43; 7th, 176; 11th, 51; 19th, 48; 22nd Guards, 226-8; (later 201st Guards), 234; 150th, 232
  Corps: Cavalry, 118, 143; I, 47-8, 51, 86, 89, 184-5; II, 47-8; III, 48, 118, 121, 123-4, 145, 268, 270-2; IV, 86, 118, 122-4, 285, 288-92; VII, 121, 124; VIII, 92, 254; X, 93, 235, 239, 241; XI, 86-9; XIII, 92-3, 220-4, 227, 229, 232, 234, 240-1; XV, 93, 276, 283, 289-92; XVII, 102, 104; XVIII, 110, 131-2, 139; XXII, 134; XXX, 220-3, 225, 230, 239-40, 260-1; XXXIII, 286, 288-9, 291-2
  Divisions: 1st Airborne, 260-2; 1st Cavalry, 22, 45-6, 118, 123; 3rd Cavalry, 54, 85, 89; 1st Armoured, 178, 192, 228, 230, 234, 241; 2nd Armoured, 215, 260; 7th Armoured, 180, 195, 214-5, 220, 229-34, 237; 10th Armoured, 240; 79th Armoured, 252, 264; Guards, 88; 1st., 22, 51, 247; 2nd, 22, 85; 3rd, 22, 95, 177; 4th, 45, 47, 51, 102; 5th, 192; 6th, 45, 52, 123; 7th, 54, 89, 95; 9th, 85, 87, 95, 102, 139; 15th, 85, 87, 135; 18th, 93, 96-7; 21st, 87; 24th, 87; 29th, 61-2, 124, 149, 151-2; 36th, 93, 96, 283; 40th, 133; 44th, 236, 238; 47th, 86; 50th, 192, 230; 51st, 120-3, 133, 192; 55th, 124, 133; 62nd, 123
  Expeditionary Force, 43-54, 59, 61, 78, 88, 138, 147, 164, 184-92
  German assessment of, 130
  Groups, Army: 11th, 278; 21st, 252-64
  Regiments and other units: 'L' Battery, R.H.A., 48; 10th Hussars, 104; Essex Yeomanry, 104; Cameronians, 170 and f.n.; 4th Royal Tank Regiment, 233; 17th Armoured Car Battalion, 143
  Council, 33, 36, 163, 175, 183-4, 193
  Canadian, 101, 109-10, 143, 251, 255-6, 258, 263-4
  Education Corps, 166

Army (contd.)
  French, 42-104 passim, 122, 127-92 passim
  Free Forces, 209, 230, 232, 249, 251
  German assessment of, 130
  German, 42-57, 67-111, 120-48, 157-60, 177-99, 191, 195, 200, 213-66 passim
  Indian, 61, 172-3, 187, 214, 265, 270, 272-3, 275, 281
    Brigades: 9th, 233; 10th, 233; 11th, 244
    Divisions: 4th, 215, 224, 227; 5th, 232, 283, 285-6, 288, 291, 292; 6th, 65; 7th, 283, 286; 12th, 274; 17th, 285, 290-2; 19th, 289-90, 292; 20th, 285; 23rd, 285
  Italian, 70, 77, 194, 216, 220, 222, 225, 231, 288
  Japanese, 206-8, 265-93
  New Zealand, 61, 187, 220, 224-5, 235, 238, 241
  Reform of, 32-6, 182-3
  Role of, 14, 175
  Russian, 56, 67, 179, 211
  South African, 187, 220, 223, 230-1, 234
  United States, 73-6, 128-9, 134, 137, 141, 145-6, 243-4, 250, 252-5, 258, 263-4, 279
    Armies: First, 145, 252, 255, 262; Third, 255; Fifth, 244-51; Seventh, 243; Ninth, 262-4
    Corps: II, 243
    Divisions: 1st Armoured, 247; 3rd 247; 82nd Airborne, 261; 101st Airborne, 261
    Group, Army: 12th, 252, 255-64
    Rangers, 247
Army Quarterly, The, 168, 170
Arnhem, 259-62
Arnold-Forster, Rt. Hon. H. O., 32, 34
Arras, 72, 76, 82, 100, 102-4, 115-6, 137, 144-5, 148, 157, 162, 191
Artillery, 15, 35-46, 52, 61, 68, 70-1, 80-126 passim, 129-30, 141-5, 157, 159, 162, 167, 174, 176, 219, 233, 235, 237, 239, 240, 251, 256, 264
Artois, 65, 84
Ashanti Expedition, 30, 265
Aspinall-Oglander, Brigadier-General C. F., 155
Asquith, H. H., Earl of Oxford and, 56, 58, 64, 66, 68, 78
Assam, 207, 278-9, 282-3
Assault, Infantry, 26, 38, 39, 42, 81, 95, 98, 109, 156

Index

Athens, 195
Atkinson, C. F., 173
Atlantic ports, 125, 209
Attack, general, *see* Offensive
Attack, limited, 73, 105, 128, 141, 145
Attrition, 25, 96, 106, 112
Aubers Ridge, 62, 81-4, 89
Auchinleck, Field Marshal Sir Claude, 170, 173, 179, 198, 201, 216, 220, 223-38, 232
Australia, 204, 273
Australian Army, *see* Army, Australian
Austria-Hungary and Army, 56-8, 60, 67, 69-70, 77
Auxiliary forces, *see* Territorial, Militia and Volunteers
Avranches, 253, 255
Axis, 187, 201, 229, 242

Baden-Powell, Colonel R. S. S., later Major-General Lord, 23
Baghdad, 68-9
Baldwin, Air Marshal Sir John, 208, 279
Balfour, A. J., Earl of, 32-3, 64
Balkans, 57-8, 60, 68, 194, 200-1, 210
Baltic ports, 211
Barclay, Brigadier C. N., 170 f.n.
Barrington-Ward, R.M., 138
Basra, 65
Battle Drill, 97-8, 173-4, 218, 273
Bazentin, 95
Beaurevoir, 118-9
Beaverbrook, Lord, 165
Beda Fomm, 195, 215
Beersheba, 74, 162
Belgian Army, 45, 54, 188
Belhammed, 230
Bengal, 275-6
Benghazi, 216, 228-9, 242
Berchtesgaden redoubt, 210
Berlin, 210
Besica Bay, 151
Biggarsberg Mountains, 21
Bir Gubi, 211, 227, 229
Bir Hacheim, 230, 232, 238
Birch, Lieutenant-General Sir Noel, 82, 92, 110, 157, 162
Birdwood, Field Marshal Lord, 140, 149, 153
Black Sea, 67
Blitzkrieg, 160, 214
Bloemfontein, 17, 22-5
Bock, Field Marshal F. von, 190
Boer,
  Republics, 13, 26, 31
  War, *see* South African War
Bourlon Ridge and Wood, 118, 120-3
Boxes, defensive, 230-1, 234
Brackenbury, General Sir Henry, 13
Bradley, General O., 211, 243, 252, 256-7, 262
Brann, Brigadier-General D. W., 245
'Break-in', 81, 83
'Break-Through', 71, 75-7, 80-3, 96, 101, 255
Breda, 259
Briand, A., 70-1
Briggs, Lieutenant-General Sir Harold, 232
Brittany, 209, 257
Brodrick, Rt. Hon. St. John, 31, 34, 41
Brooke, *see* Alanbrooke
Brooke-Popham, Air Chief Marshal Sir Robert, 266, 274
Broodseinde, 110, 140
Brusilov, General, 67
Bulair, 149, 151
Bulgaria and Army, 58, 63, 67-8, 76-7
Buller, General Sir Redvers, 16-20, 22-3, 25
Bullercourt, 141
Bulow, Field Marshal K. von, 47, 54
Burma,
  campaign, 173, 205, 211, 229, 265, 273, 275-93
  Corps, 275
  Road, 205-6, 275-6
Burnett-Stuart, General Sir John, 177
Buthidaung, 283
Butler, Lieutenant-General Sir William, 17-8
Butterworth, 270
Byng, Field Marshal Viscount, 100-1, 104, 118-9, 121, 161

Cabinet, the British, 17-8, 20, 56, 59, 65, 71-2, 105, 124, 137-8, 164, 182, 185, 196
Cadorna, General L., 70, 74
Cadre training system, 174
Caen, 253-4
Calais, 47, 54, 64-5, 72, 256-7, 259
Calcutta, 275-6, 283
Calvert, Brigadier J. M., 277
Cambrai, 74, 76, 102, 113-27, 130, 138, 140, 142, 159-61, 191
Cambridge, H.R.H. Duke of, 13, 30, 41
Campoleone, 246-8
Canal du Nord, 118
Canadian Army, *see* Army, Canadian
Cape Colony, 20, 22
Cape Town, 17, 19, 21
Caporetto, 70, 74

304 British Generalship in the Twentieth Century

Carton de Wiart, Lieutenant-General Sir Adrian, 102
Cassino, 244, 249, 251, 257
Casualties, details of, 93, 99-100, 104, 108 f.n., 110, 112, 120-22, 214, 223, 256, 264, 278, 288
Cauldron, the, 232-3, 236
Cavalry, 24, 26-7, 38-9, 42, 46-7, 51-2, 55, 84-5, 88-90, 95, 102, 104, 110, 118, 137-8, 143, 157, 219, 240
 charge, 26, 27, 30, 37-8
Ceylon, 205-7,
Chamberlain, Rt. Hon. Neville, 187
Champagne, 65, 84
Channel ports, 54, 72, 76, 105, 136, 145, 190, 262
'Charing Cross', 241
Chemin des Dames, 50, 52, 96, 121, 134
Cherbourg, 192, 253
Chetwode, Field Marshal Lord, 100, 173
Chiang Kai Shek, 206, 279
Chiefs of Staff,
 British, 33, 163-4, 196-9, 207, 266, 289
 U.S., 197, 201, 204-7
 Combined, 197, 204-7
China and Army, 205-6, 275-9
Chindits, 265-8, 280, 285-8
Chindwin River, 207, 275-8, 282-3, 288-9
Chittagong, 277
Chunuk Bair, 153
Churchill, Rt. Hon. Sir Winston, 56-8, 62, 114, 187-212, *passim,* 229, 234, 272, 277, 287, 294-5
Cisterna, 246-8
Citizen Army, 31, 35, 99, 146
Clark, General M. W., 244-51
Clausewitz, General Carl von, 166
Cleave, 264
Clery, Lieutenant-General Sir Francis, 29
Cole, Brigadier D. H., 166
Colenso, 20, 23, 25
Colesberg, 22
Collingwood, Lieutenant-General Sir George, 170 f.n.
Combined operations, *see* Amphibious Operations
Combined Training, 37
Command, Unified Allied, 76
Commander, position in battle, 40, 86, 104, 122, 142, 158, 182
Commander, qualities of and relationship to subordinates, 152-3, 155-6, 162
Commandos, 264, 291

Communications, 81, 86, 89, 122, 124, 130, 153, 158, 290
Compiègne, 48
Congreve, Lieutenant-General Sir Walter, 93
Coningham, Air Marshal Sir Arthur, 227
Conscription, *see* National Service
Constantinople, 61, 63, 65
Continental strategy and diplomacy, 31, 33, 36, 41, 59, 171
Co-ordination of arms, 114, 126, 178, 215, 218-9, 224, 233, 240
Coral Sea, battle of, 205, 276
Counter-attack, 26, 37, 40, 74, 101, 108-9, 111, 125, 131, 139, 142, 144, 191-2, 233, 248, 262-3, 270-1
Counter-offensive, 47-9, 76, 121, 124, 128-9, 134, 259, 262-3
Courtrai, 145
Cowan, Major-General D. T., 285
Crete, 216 ,
Crimean War, 13, 15
Croft, Brigadier-General W. D., 102
Cronje, General P., 24
Cruewell, Lieutenant-General L., 218, 223-5, 228
'Crusader' Offensive, 216-7, 219-26, 237
Ctesiphon, battle of, 65
Cunningham, Admiral of the Fleet Viscount, 197
Cunningham, General Sir Alan, 219-25
Cyrenaica, 194
Czechoslovakia, 211

Damascus, 75, 77
Dardanelles, *see* Gallipoli
 Committee, *see* War Council
Dawnay, Major-General Guy, 155
Defence, 26, 38-40, 46, 55, 73, 76, 83, 87-90, 100-1, 111, 116, 124-5, 128-33, 136, 138, 149, 151, 162, 167, 188, 192, 195, 213, 263, 287
 Committee, 55, 196, 197, 198
 Ministry of, 33
de Fonblanque, Major-General E. B., 181
de Guingand, Major-General Sir Francis, 238, 258
Delville Wood, 95
Dempsey, Brigadier M., later Lieutenant-General Sir Miles, 214, 252, 260, 262
Denmark, 211
Derna, 229, 236
de Robeck, Vice-Admiral J. M., 62

Index

Desert War, 173, 176, 178-9, 181, 201, 213-42
Deverell, Field Marshal Sir Cyril, 184
De Wet, General C., 24-5
Dill, Field Marshal Sir John, 170, 179, 184-5, 195, 197, 211
Dimapur, 286
Disarmament, 171
Discipline, 42, 161, 281
Djebel, 229
Dobbie, Lieutenant-General Sir William, 266
Dojran, 77
Douai, 84, 102, 118, 190
Doullens Conference, 136
'Dragoon', *see* 'Anvil'
Drakensberg Mountains, 17
Dugan, Major-General Lord, 139, 173
Dundee, 20
Dunkirk, 54, 189, 192-3, 225
Durban, 18, 20, 22
Dyle River, 188-9, 225

Easterners *versus* Westerners, 56-76
East London, 17, 22
Eden, Rt. Hon. Anthony, later Lord Avon, 195
Edmonds, Brigadier-General Sir James, 93 f.n., 112, 132, 160
Education, military, 27-8, 37, 166
Egypt, 60-2, 69, 142, 148, 180, 185, 187, 194-5, 229, 237, 265
Eindhoven, 201
Eisenhower, General of the Armies D. D., 198, 201, 210-12, 243, 248, 252-64 *passim*
Elansaagte, battle of, 21
Elbe, River, 210-1
Elgin Commission, 24, 26, 29, 31-2
Elles, General Sir Hugh, 113, 115
Ellison, Lieutenant-General Sir Gerald, 33
Encyclopaedia Britannica, 173
Engineers, 85, 100, 112, 115, 139, 264
Entrenchments, 26, 37-8, 40, 46, 54-5, 60, 80, 82-3, 114, 119, 157, 213, 230-1
Eritrea, 194
Esher, Rt. Hon. Viscount, 32-3, 36, 183
Estcourt, 18, 20
Evetts, Lieutenant-General Sir John, 139, 170 f.n., 173
Expeditionary Force, 13, 15, 17, 35-6, 187, *see also* Army, British
Experimental Force, 176-7

Exploitation of success, 42, 81-2, 93, 95, 105, 125

Faisal Ibn Husain, Prince, later King of Iraq, 69
Falaise, 253-5
Falls, Captain Cyril, 99 f.n.
Far East, 187, 204-8
Fergusson, Brigadier Sir Bernard, 277
Festubert, 83, 138
Field Service Regulations, 39-43, 52, 166, 175
Fighting Forces, 168
Financial stringency, 16, 171, 175-6, 178-9
Fire plan, 174
Flesquières, 120-124
Flying Corps, Royal, 111, 119, 158 f.n.
Foch, Marshal F., 74, 76, 84, 86, 88, 104, 128, 133-8, 145, 147-8, 166
Formosa, 204
France and French Policy, 33, 59, 61, 63-4, 84, 127, 186, 188, 191, 201, 209-11
France, battle of, 189-92
Franchet d'Espèrey, General, 76
Franklyn, General Sir Harold, 214
French, Field Marshal Sir John, Earl of Ypres, 22, 24, 27, 45, 47-8, 52-3, 56, 60-1, 65, 83-8
French Army, *see* Army, French
Freyberg, Lieutenant-General Lord, 220
Frost, Major-General J. D., 170 f.n., 261
Fuller, Major-General J. F. C., 96, 115, 158, 161, 163, 166-8, 170, 177, 180

Gaba Tepe, 149, 153
Gabr Saleh, 222
Gaeta, 244
Galicia, 56, 67
Galliéni, General J. S., 49, 54, 58
Gallipoli, 51, 57-9, 61-6, 140, 148-56, 162
Galloway, Lieutenant-General Sir Alexander, 170 f.n., 223-5
Gambut, 222
Gamelin, General M. G., 188, 191
Garigliano, R., 244
Gas, 82, 85, 95
Gatacre, Major-General Sir William, 22, 29
Gatehouse, Major-General A. H., 240

Gaza, 69, 72, 74, 162
Gazala, 227, 229
  battle of, 201, 216-7, 228, 230-7
General Staff, 33, 36, 55, 59, 63-4, 67, 70, 74, 78, 84, 106
'George', German plan, 129
Georges, General, 188
German Army, *see* Army, German
Germany, 31, 64, 67, 208, 210-1, 265-6
General Headquarters (G.H.Q.), 47, 59, 78, 90-1, 138-9, 191
Gheluvelt, 106, 108-9
Giffard, General Sir George, 198, 278-9, 282, 289, 293
Glencoe, 20
Godwin-Austen, General Sir Alfred, 220, 223-4, 229
'Goodwood' (fighting around Caen), 254-5, 257
Gort, Field Marshal Viscount, 173, 182, 184-5, 189, 191-3
Gott, Lieutenant-General W. H. E., 220, 229, 236
Gough, General Sir Hubert, 96, 105-6, 109, 135
Government, *see* Statesmen and Cabinet
Gracey, General Sir Douglas, 173
Graincourt, 123
Graham, Major-General D. A. H., 170 f.n.
Grave, 260-1
Great War, the (1914-18), 13, 25, 43, 45-162, 188, 213, 236, 240, 265, 294
Greece and Army, 57-8, 61, 63-4, 77, 194-5, 216
Grey, Sir Edward, later Rt. Hon. Earl, 64
Grigg, Rt. Hon. Sir James, 199
Ground, tactical, 50, 52, 90, 124, 138
Guerrilla operation, 25, 69, 265, 277
Guggisberg, Brigadier-General Sir Gordon, 139
'Gustav Line', 244
Gwynn, Major-General Sir Charles, 170

'Hagen', operation, 134
Haig, Field Marshal Earl, 25, 27, 35, 43, 47, 51, 53-4, 65, 69, 71,-149 *passim*, 156-62, 198, 294
Ham, 143
Hamel, battle of, 138, 140, 142-3
Hamilton, General Sir Ian, 27, 37, 59-60, 62, 147-56
Hankey, Rt. Hon. Lord, 60, 71, 79, 164, 197

Harding, Field Marshal Lord, 241
Harper, Lieutenant-General Sir George, 121
Hartington, Marquis of, later Duke of Devonshire, 32
Hasted, Major-General W. F., 282
Haugh, Major-General J. W. N., 170 f.n.
Hawes, Major-General L. A., 186
Hazebrouck, 76, 129, 134
Heath, Lieutenant-General Sir Lewis, 268
Hejaz Revolt, 69
Helles, Cape, 62, 149-53
Henderson, Colonel G. F. R., 25-6, 37, 39, 42, 164, 182
Heneker, General Sir William, 132
High explosive shells, 81, 95, 102, 139, 178-9
High Wood, 95
Hill 70, 109-10
Hindenburg, Field Marshal P. von, 56, 69
Hindenburg Line, 72, 102, 116, 118-21, 138, 145
History, Military, 27, 163-5
Hitler, Adolf, 194, 199, 202, 238, 255-6
Hobart, Major-General Sir Percy, 179, 181, 214, 252
Holland, 56, 188-9, 211
Holland, Lieutenant-General Sir Arthur, 101, 116
Hollis, General Sir Leslie, 196
Hollond, Major-General S. E., 139
Home Defence, 14, 34, 193
Hopin, 287
Hore-Belisha, Rt. Hon. Lord, 182-4
Horrocks, General Sir Brian, 214, 240, 264
Horses and Horsemanship, 42, 167, 181
'Hump' Route, 205, 207
Hunter, Lieutenant-General Sir Archibald, 19
Hunter-Weston, Lieutenant-General Sir Aylmer, 51, 149-153

Imperial Defence, 14, 31, 36, 54, 171-2, 175, 186-7
Imperial Military Geography, 166
  Committee of, 33, 55, 164
Imphal-Kohima, battle of, 208, 282-3, 285-8
Indaw, 285
India, 14, 18-20, 33, 36-7, 54, 65, 172, 175, 187, 205, 207, 265, 275-8, 282
Indian Army, *see* Army, Indian
Indirect approach, theory of, 160

Index

Industrial expansion and production, 178-9, 187, 213, 295-6
Infantry, 46, 50, 68, 81, 84-5, 90-114, 123-4, 130-3, 140-5, 157, 159, 167, 173, 176-7, 232, 239, 251, 264
  Tactical formations of, 37, 81, 91, 93, 102, 109-12, 119-20, 130, 139
Infantry Drill Book, 37
Infantry Training, 173
Infiltration, 130, 167
Initiative, military, 16, 19, 22-3, 25-6, 39, 43, 98, 137, 149, 152, 167, 172, 194, 207, 222, 224, 237, 253, 268-9, 271, 275, 280, 296
Intelligence, military, 28-9, 102, 186, 191, 260, 280
Invasion scare, 13, 45, 193
Irrawaddy, River, 207, 277-81,
Ironside, Field Marshal Lord, 170, 184-5
Irwin, Lieutenant-General N. M. S., 276-7
Ismay, General Lord, 164, 196
Italian Army, *see* Army, Italian
  campaign, 202-4, 212, 214, 243-52, 264, 268, 273
Italy, 57, 69, 70, 74, 127, 186, 188, 194-5, 200, 202, 204, 208, 210

Japan, 187, 199, 204, 211, 229
Japanese Army, *see* Army, Japanese
Jerusalem, 74-5
Joffre, Marshal J. J. C., 46-8, 54, 56, 69, 71, 82-5, 91, 100, 188
Johore, 268, 270, 272, 274
Jomini, Baron, 166
Julian Alps, 70
Jungle Warfare Training, 272-3, 278

Kabaw Valley, 288
Kaiser William II, 76
Kalewa, 288-91
Katamura, Lieutenant-General S., 289
Kawabe, Lieutenant-General H., 289
Kemal, Mustapha, 154
Kesselring, Field Marshal, 244-5
Kiggell, Lieutenant-General Sir Lancelot, 161
Kimberley, 21-24
Kimura, Lieutenant-General H., 289, 290
King, Fleet Admiral E., 204-5
Kirke, Lieutenant-General Sir Walter, 176
Kitchener, Field Marshal Earl, 15, 23-4, 29, 36-7, 48, 55-60, 62-7, 79, 83, 85, 148, 163
Klopper, Major-General H. B., 234
Kluang, 271

Kluck, General A. von, 46-50
Kluge, Field Marshal G. von, 255
'Knightsbridge', North Africa, 230, 234
Knox, General Sir Harry, 184
Kohima, *see* Imphal
Kota Bahru, 266-8, 270
Kra Isthmus, 206, 268
Kruger, P., President of the Transvaal, 18, 24
Kuala Lumpur, 271
Kuantan, 266, 272
Kut, 65, 68

La Bassée, 84, 138
Ladysmith, 17, 19, 20-1, 23-4
La Fère, 76
Laing's Nek, 17
'Lancashire Landing', Gallipoli, 151
Lance, 26-7
Landing craft, 200, 202, 204-5, 244, 252, 290
Lansdowne, Rt. Hon. Marquis of, 15, 17-20, 23, 29, 31
Lashio, 275-7
Lawrence, T. E., 69, 162, 265
Leadership, 26, 110, 138, 155, 161, 281, 287, 295
Le Cateau, 45-6
Ledo, 205, 276, 278, 285
Leese, Lieutenant-General Sir Oliver, 198, 239, 289, 291
Lemnos, 61
Lens, 84, 87, 109
Lentaigne, Major-General W. D. A., 288
Liddell Hart, Captain Sir Basil, 99 f.n., 112 f.n., 132, 161, 163, 166-8, 173, 183-4
Lille, 45
Liri Valley, 249
Ljubljana Gap, 70, 209
Lloyd George, Rt. Hon. David, 57, 60-1, 67-75, 112, 127, 148, 164, 197
London, defence of, 14, 260
Longueval, 95
Loos, 65, 84-9
Lubeck, 211
Lucas, Lieutenant-General J. P., 244-9
Ludendorff, General E., 56, 69, 76, 129, 133-4, 147
Lumsden, Lieutenant-General Sir Herbert, 239
Lys, River, 76, 106, 133, 135-7, 145

MacArthur, General of the Armies D., 204
Macedonia, 59, 63, 65, 67, 76

Machine gun, 15, 39-40, 43, 46, 60, 80, 93, 100-1, 111-2, 114, 126, 130, 132-3, 138-9, 157, 178
Mackensen, General E. von, 247
Mackensen, Field Marshal A. von, 67
Mafeking, 21, 23
Magersfontein, 22, 26
Maginot Line, 188
Malacca, 271
Malaya, 25, 205-6, 265-74, 291
Malta, 19, 229, 235
Mandalay, 277, 289, 290-1
Mangin, General, 70
Manoeuvre, 25-6, 147, 166
Manoeuvres, 28, 41, 193
Manoury, General M.-J., 51
Marcoing, 118, 120
Mareth Line, 242
Marines, Royal, 52, 264
Marlborough, Duke of, 147, 156, 174, 212
Marne, First battle of the, 49-51, 101,
Second battle of the, 76, 134, 138,
Marriott, Major-General Sir John, 227
Marseilles, 262
Marshall, Major-General F. J., 139
Marshall, General of the Armies G. C., 201, 211
Masnières, 118, 120
' Matador ' plan, 268
Maubeuge, 45-6, 145
Maude, Lieutenant-General Sir Frederick, 68-9
Maungdaw, 283
Maurice, Major-General Sir Frederick, 163-5
Maurice, Major-General Sir J., 164
Maxse, General Sir Ivor, 43, 93, 96-7. 104, 110, 125, 130-2, 139, 159, 173-4
Mayu Peninsula, 227, 283
Mechanization, 172, 175-6, 179, 251, 290
Mechili, 229
Medina, 69
Mediterranean, 59-60, 64-5, 194, 200, 206, 208-9
Megiddo, 76-7, 162
Meiktila, battle of, 181, 290-2
Menin, 54, 109
Mersa Matruh, 235, 241
Mersing, 272
Mesopotamia, 75
Messervey, General Sir Francis, 173, 181, 232, 289
Messines Ridge, 73, 84, 91, 104-6, 109, 140
Meteorology, 116

Methuen, Field Marshal Lord, 22, 29
Metz, 257
Meuse, River, 45, 190, 262-3
Mézières, 145
' Michael ' plan, 129
Middle East, 65, 187, 226, 273
Midway Island, 199, 205, 276
Militia, 15, 23, 28, 34-5
Miller, Major-General C. H., 227
Milne, Field Marshal Lord, 68, 175-6
Milner, Rt. Hon. Viscount, 18
Minefields, 231-2, 238-9, 252
Mobility, 26, 138-9, 142-4, 162, 167
Mobilization, 13, 16, 28-30, 45
Modder, River, 24
Model, General, later Field Marshal W., 261
Mogaung, 287
Moltke, General H. J. L. von, 149
Momentum of the attack, 142, 176
Monash, General Sir John, 140-1, 159
Monchy-le-Preux, 102
Money, Major-General R. C., 157 f.n., 170 f.n.
Mons, retreat from, 46-9, 52-3
Monro, General Sir Charles, 64, 91
Montauban, 90
Montgomery, Field Marshal Viscount, 79, 170, 173, 179-82, 198, 202, 210-1, 214, 219, 225, 236-44, 248, 252-63, 295
Montgomery-Massingberd, Field Marshal Sir Archibald, 161, 176
Mont St. Quentin, 144
Morale, 49-50, 69, 83, 90, 100, 145, 149, 161, 238, 271, 279-81, 285, 292
Morgan, Lieutenant-General Sir Frederick, 173
Mountbatten, Admiral of the Fleet, Earl, 181, 198, 207, 252, 276, 278, 289, 293
Mounted Infantry, 27, 149
Muar, 271-2
' Mulberry ' Harbour, 252
Munich Crisis, 178
Munitions, Ministry of, 84
Murray, General Sir Archibald, 63, 65-6, 69, 72-3
Murray, General Sir Horatius, 170 f. n.
Musketry, 41-2, 46, 55
Mussolini, B., 202, 238
Mutaguchi, Lieutenant-General R., 288-9
Myingyan, 291
Myitkyina, 277-8, 283, 286-7

Index

Naauwpoort, 22
Namur, 45, 84
Napoleon I, Emperor of the French, 27, 106, 295
Narvik, 189
Natal, 17-20, 22-3
National Service, 34
Navy, 14, 31, 58, 60-2, 148-9, 153, 189, 192, 217, 229, 264, 266, 269
Naval Division, the Royal, 54
Neame, Lieutenant-General Sir Philip, 195, 242
Neuve Chapelle, 62, 80-3, 115
New Armies, 55, 60-1, 65, 85, 87, 91, 153-4
New Zealand Army, see Army, New Zealand
Night advance and attack, 55, 93, 95, 155, 176, 233, 239
Nijmegen, 260-1, 263
Nile River, 201, 234
Nimitz, Fleet Admiral C. W., 204
Nivelle, General R. G., 52, 71-2, 76, 100, 104-5, 148, 160, 165
Nixon, General Sir John, 65
Normandy, battle for, 252-7, 264
Noyelles, 118
North Africa, 201, 252, 266
Northern Area Combat Command, 279
North West Europe, 194-5, 208-12, 252-64
Norrie, Lieutenant-General Lord, 220, 222, 225, 235
Norway, 189
Nye, General Sir Archibald, 170, 199

O'Connor, General Sir Richard, 170 f.n., 173, 178-80, 194-5, 214-5, 219, 229, 237, 295
Oder River, 210
Offensive, 26-7, 38-40, 76, 80-4, 97, 111, 113, 116, 126, 129-30, 132, 135, 138, 140-4, 148, 159-60, 188, 235, 265, 276, 282
  German, March-April, 1918, 131-8
Officer, role of, 27, 124, 125, 132
Oilfields, Anglo-Persian, 57, 65
Old, Brigadier-General W. D., 279
Omdurman, 15
Operations, Military, Director of, 59, 164, 196
Orange Free State, 16, 17, 20, 22
Orange River Bridge, 21-3, 25
Orne, River, 253, 255
Ostend, 45, 56, 106
Oudenarde, 147
'Overlord', 180, 199-200, 202, 204-6, 208, 214, 241, 252

Paardeberg, 24
Pacific, 204, 208, 211-2, 276
Paget, General Sir Bernard, 170, 173-4, 179
Palestine, 69, 72, 104, 162, 167, 187, 194, 225, 265, 277
Pakokku, 290
Panzer Forces, 166, 215-235 *passim*
Parachute troops, 290; *see also* Arnhem
Paris, 48-9, 61, 76, 136-7, 191, 257
Parliament, 34, 58, 66, 164, 294
Passchendaele, *see* Ypres, third battle of
Ridge, 105-6, 110
Patrols, 25, 248
Patton, General G. S., 211, 243, 255
Pegu, 292
Peiwar Kotal, 15
Penang, 270
Percival, Lieutenant-General A. E., 266, 269-74
Péronne, 144
Perrowne, Major-General L. E. C. M., 288
Pershing, General of the Armies J. J., 128, 141, 145
Pétain, Marshal, 73, 76, 104, 127-8, 134-8, 145
Philippines, 202, 206
'Phoney War', 187, 191
Pietermaritzburg, 20
Pilckem Ridge, 108
Pillboxes (concrete), 132, 188
Plan XVII, 46
Plans and Planning Staff, 33, 153, 156, 196, 202, 238, 283, 286, 292
Playfair, Major-General I. S. O., 181
Plumer, Field Marshal Viscount, 91, 96-7, 104-6, 109
'Pluto' (the Pipe Line Under The Ocean), 252
Poetry and War Poets, 100
Poland, 177, 188, 195, 208
Polygon Wood, 105
Poplar Grove, 24
Port Elizabeth, 17
Portuguese Army, 133, 136
Pownall, Lieutenant-General Sir Henry, 274
Prague, 211
Pretoria, 17, 22, 24-5, 31
Press, 253, 263, 278
Prince of Wales, H.M.S., 269
Principles of War, 43, 165-6
Prome, 292
Prussia (East), 56

Pulteney, Lieutenant-General Sir William, 124
Pursuit, 34, 39, 74, 202, 235, 241, 257, 260-1, 282, 291

Quattara Depression, 235
Queen Elizabeth, H.M.S., 153

Radar, 252
Raids, 26, 248, 290
Ramree Island, 206, 290
Ramsay, Admiral Sir Bertram, 192
Rangoon, 206-7, 275-6, 282, 289, 291-2
Ranking, Major-General R. P. L., 286
Rawlinson, General Lord, 37, 89-91, 93, 95, 105, 135, 141-3, 161, 173
Reconnaissance, 38, 42, 47, 84, 155, 174-5, 182, 247
Regular Army, 14, 26, 31, 33-6, 55, 88, 146, 214
Reichswald, 263
Reinforcements, 38-9, 44, 98, 100, 131, 137, 228, 282
Reims, 47, 69, 135, 145
Remagen Bridge, 264
Repulse, H.M.S., 269
Reserves, 54-5, 60, 71, 81-3, 86-9, 109, 127, 129, 131, 134-7, 142, 148-9, 152, 158, 162, 256, 270-1, 280, 283, 285
Rhine, River, 208-10, 257, 260, 263
Rhineland, battle for the, 211, 262-4
Rhodes, Cecil, 21
Riddell-Webster, General Sir Thomas, 170
Rifle, 15, 25-7, 46, 110, 126
Ritchie, General Sir Neil, 225-35
River Clyde, S.S., 151
Roberts, Field Marshal Earl, 13, 15, 21, 23-5, 27-30, 32, 265, 294
Robertson, Field Marshal Sir William, 65-6, 69, 71, 75-6, 78-9, 112, 127-8, 148, 163, 165, 196, 198
Rome, 70, 208-9, 244-5
 battle for, 247, 249-51
Rommel, Field Marshal E., 169, 178, 192, 215-42, *passim*, 255-6
Roosevelt, F. D., President of U.S.A., 199, 201, 210-1, 277
Roulers, 106
Ruhr, 208, 210, 257, 259
Rumania, 57, 67-8
Rundstedt, Field Marshal G. von, 190, 255-6
R.U.S.I. Journal, 160, 168

Russia, 33, 56, 60, 65, 66, 73, 75, 84, 128, 193, 199, 210
Russian Army, *see* Army, Russian
Revolution, 73
Russo-Finnish War, 186
Russo-Japanese War, 33, 37
Ruweisat, 236

Saar, 192
St. Mihiel, 145
St. Nazaire, 192
St. Quentin, 72, 76, 118, 127, 129
Salerno, 202, 245
Salisbury, Rt. Hon. the Marquis of, 19
Salonica, 57, 60, 63-4, 68, 73, 288
Sambre, River, 46
Sardinia, 202
Sarrail, General, 63, 68
Sato, Lieutenant-General, 286
Scarpe, River, 102, 104
Scheldt, River, 138, 259-60
Schwaben Redoubt, 97
Scoones, General Sir Geoffrey, 285
Sea, command of, 57, 65, 269, 270, 273
Sedan, 190
Seine, River, 253
Sensée, River, 118
Serbia, 57-8, 61, 63-8, 77
Shipping, 19, 31, 72, 189, 192, 200, 205, 207, 243
Shrapnel, 81, 102
Shwebo, 289, 291
Siam, 268-9
Sicily, 202, 243
Sidi Omar, 224
Sidi Rezegh, 220, 222-3
Siegfried Line, 258, 263
Signals, *see* Communications
Signora, 268
Sinai, 69
Singapore, 205, 229, 265-8, 270, 274-5, 289
Sittang, 289
 River, 275
Skopje, 77
Slim, Field Marshal Viscount, 170, 173, 179, 192, 198, 207, 275-93, 295
Slim, River, 271
Smith, Lieutenant-General Sir Arthur, 226
Smith-Dorrien, General Sir Horace, 47, 53
Smoke, 102, 111, 139, 179
Smuts, Field Marshal, Rt. Hon. J. C., 124
Snelling, Major-General A. H. J., 282

Index

Snow, Lieutenant-General Sir Thomas, 124
Soissons, 48, 51, 144
Somme, battle of, 67, 69, 70-1, 73, 80, 90-100, 101, 104, 108, 114, 157-61
Somme, River, 76, 91, 133, 143-5, 178, 190
South African Army, see Army, South African
War, 13-32, 37, 42, 53, 156, 265, 294
South Beveland, 259
South East Asia, 205-8, 252, 274
Spears, Major-General Sir Edward, 72 f.n., 165
Special Reserve, 35
Staff and Staff Colleges, 27-8, 33, 36-7, 43, 59, 87, 104, 115, 158-9, 161, 164, 166, 169, 170-1, 180, 186, 224, 237, 259, 264, 296
Stalin, Generalissimo Joseph S., 210
Stalingrad, 199
Stanhope Memorandum, 14-5
Statesmen and Soldiers, relations between, 16, 28, 33, 35, 58, 66-7, 70, 75, 126, 148, 163, 198, 211, 220, 268, 270-2, 294
Steyn, M., President of the Orange Free State, 24
Stilwell, General J. W., 275, 278-9, 285-7
Stopford, Lieutenant-General Sir Frederick, 154
Stopford, Lieutenant-General, Sir Montagu, 288
Stormberg, 22
Storm Troops, German, 130-1
Stratemeyer, Major-General G. E., 279
Stronghold, Concept of jungle fighting, 287
Student, General K., 261
Sudan, 194, 215, 265
Suez Canal, 57, 69, 186
Sumatra, 206, 289
Supreme Allied War Council, 74, 127-8, 133
Supreme Headquarters Allied Expeditionary Force (SHAEF), 260-1
Surprise and Deception, 26, 37-9, 74, 81-2, 87, 90, 95, 101, 108, 114, 116, 119, 123, 129, 130, 142, 148-9, 153, 157, 159, 162, 167, 176, 213-4, 239, 252, 254, 256, 292
Survey, 115-6
Suvla Bay, 63, 153, 155
Swinton, Major-General Sir Ernest, 80, 96, 113-5, 158
Sword, cavalry maintenance of, 26-7, 42

Tactical Lessons, doctrine and theory, 25, 27, 53, 88, 90, 93, 96-9, 101, 109, 112, 115, 121, 125-6, 159, 166, 182, 213, 218, 258, 264, 280-1, 294
Talana, 20
Tamu, 285, 288
Tanks, 74, 80, 82, 95-6, 108-9, 111, 113-23, 138, 141-3, 158-9, 161, 167, 175-9, 213, 216, 220, 230, 232-4, 237-41, 251, 254-5, 277, 279-80, 283, 289, 291-2
American, 179, 237
Drill and tactics of, 119-21, 159
Tannenberg, 56
Tedder, Marshal of R.A.F., Lord, 254
Teheran Conference, 208
Tel-el-Kebir, 15
Territorial Force and Army, 35, 55, 59, 61, 174, 183, 187
Thiepval, 93, 96-7
Tiddim, 285, 288
Times, The, 139, 164, 168
Tmimi, 236
Tobruk, 21, 194-5, 216, 220-4, 226-31, 234-5
'Torch', landings in North Africa, 201-2, 242
Tokyo, 206
Tournai, 45
Tower Hamlets, 109-10
Townsend, Major-General C. V. F., 65
Traffic control, 86, 264
Training, 37, 41-3, 69, 82, 84, 93, 98, 110, 119, 124-5, 129-31, 139, 141, 143, 159, 162, 171-4, 176, 191, 214, 220, 257, 273-4, 287, 292, 294
Transport, 16, 28-9, 31
Transvaal, 16, 19-20, 23
Trenches, see Entrenchments
Trench Warfare, definition of the problem of, 157
Trentino, 70
Trig Capuzzo, 230, 232
Tripoli, 194-5
Trondheim, 189
Tropical Diseases, 275, 280-1
Truscott, Lieutenant-General L. K., 250
Tudor, Major-General Sir Hugh, 95, 102
Tugela, River, 18, 20

Tunisia, 201-2
Turkey, 56-8, 60, 64-5, 75, 194
Turkish Army, 60, 68-9, 75, 77, 148-9, 151, 162, 167
Tyrol, 70

Uniacke, Lieutenant-General Sir Herbert, 138-9
United States, 73, 199-212
  Air Force, 207, 279
  Army, *see* Army, United States
  Marines, 204
  Navy, 204, 266

Valenciennes, 46
Valmontone, 247, 250-1
Van Reenen's Pass, 20
Vardar, River, 77
Venetia, 70
Venizelos, M., 58, 63
Verdun, 49, 54, 67, 84, 91, 96, 100-1, 104, 129, 145, 160
Victoria Cross, 30
Vienna, 70, 209
Vimy Ridge, 72, 87, 100, 105, 110
Vistula, River, 56
Vittorio Veneto, 77
Volunteers, 14-5, 23, 28, 34

Walcheren, 259-60
Wanting, 275
War Cabinet, 72, 74, 105, 124, 137, 138, 164-96
  Council and Committee, 56, 58-64, 66
  Office, 13, 32-5, 37, 55, 59, 61, 63-4, 68, 113, 173, 183, 186, 192, 196
Warren, Lieutenant-General Sir Charles, 29
Warsaw, 208
Washington, 199, 202
Waterloo, 13, 147, 163
Watter, Lieutenant-General F. von, 122
Wavell, Field Marshal Earl, 168, 179-80, 194-5, 197, 215-7, 226, 237, 274-7, 295
Weather, 73, 83, 85, 89, 92, 96, 108-10, 249, 251, 255, 261, 263, 275, 279, 288, 291

Wellington, Duke of, 13, 147, 156, 174, 295
Wesel, 260, 263-4
Westphal, Lieutenant-General S., 218, 225
Wetzell, Lieutenant-Colonel, 129, 133
Weygand, General M., 74, 190, 192
White, General Sir George, 19-23, 26, 29-30, 32
Whiteley, General Sir John, 223
Will of the Commander, 49, 149, 155
Williams, General Sir Guy, 181
Williams, Major-General W. de L., 132
Wilson, Field Marshal Lord, 179, 198
Wilson, Field Marshal Sir Henry, 35, 37, 74-5, 148, 198
Windsor, Duke of, 191
Wingate, Major-General O. C., 207, 277, 283
Wire entanglements, 25, 37, 55, 60, 80-2, 87, 90, 95, 98, 102, 110, 114-6, 119-20, 126, 132, 230-1, 238
Wireless, 82, 122, 158
Withdrawal, 46-49, 55, 72, 124, 132, 138, 270-1, 280, 282, 285, 289
Wolfe Murray, Lieutenant-General Sir James, 59, 62
Wolseley, Field Marshal Viscount, 13-25, 27-9, 32, 164, 265, 294
World War, Second, 79, 163, 176, 186-296

Yalta Conference, 210
Yennenyaung Oilfields, 290
Yeomanry, 15, 35
Ypres, 54, 56, 72, 82, 83, 105, 124, 145
  Third battle of, 73-4, 105-12, 113, 127, 140, 142, 159-60, 164
Yugoslavia, 194
Yunnan, 275

Zeebrugge, 54, 56
Zululand, 265